WHAT EVERY ENGINEER SHOULD KNOW ABOUT

PATENTS

WHAT EVERY ENGINEER SHOULD KNOW

A Series

Editor

William H. Middendorf

Department of Electrical and Computer Engineering
University of Cincinnati
Cincinnati, Ohio

Vol. 1 What Every Engineer Should Know About Patents, *William G. Konold, Bruce Tittel, Donald F. Frei, and David S. Stallard*

Vol. 2 What Every Engineer Should Know About Product Liability, *James F. Thorpe and William H. Middendorf*

Vol. 3 What Every Engineer Should Know About Microcomputers: Hardware/Software Design: A Step-by-Step Example, *William S. Bennett and Carl F. Evert, Jr.*

Vol. 4 What Every Engineer Should Know About Economic Decision Analysis, *Dean S. Shupe*

Vol. 5 What Every Engineer Should Know About Human Resources Management, *Desmond D. Martin and Richard L. Shell*

Vol. 6 What Every Engineer Should Know About Manufacturing Cost Estimating, *Eric M. Malstrom*

Vol. 7 What Every Engineer Should Know About Inventing, *William H. Middendorf*

Vol. 8 What Every Engineer Should Know About Technology Transfer and Innovation, *Louis N. Mogavero and Robert S. Shane*

Vol. 9 What Every Engineer Should Know About Project Management, *Arnold M. Ruskin and W. Eugene Estes*

Vol. 10 What Every Engineer Should Know About Computer-Aided Design and Computer-Aided Manufacturing: The CAD/CAM Revolution, *John K. Krouse*

Vol. 11 What Every Engineer Should Know About Robots, *Maurice I. Zeldman*

Vol. 12 What Every Engineer Should Know About Microcomputer Systems Design and Debugging, *Bill Wray and Bill Crawford*

Vol. 13 What Every Engineer Should Know About Engineering Information Resources, *Margaret T. Schenk and James K. Webster*

Vol. 14 What Every Engineer Should Know About Microcomputer Program Design, *Keith R. Wehmeyer*

Vol. 15 What Every Engineer Should Know About Computer Modeling and Simulation, *Don M. Ingels*

Vol. 16 What Every Engineer Should Know About Engineering Workstations, *Justin E. Harlow III*

Vol. 17 What Every Engineer Should Know About Practical CAD/CAM Applications, *John Stark*

Vol. 18 What Every Engineer Should Know About Threaded Fasteners: Materials and Design, *Alexander Blake*

Vol. 19 What Every Engineer Should Know About Data Communications, *Carl Stephen Clifton*

Vol. 20 What Every Engineer Should Know About Material and Component Failure, Failure Analysis, and Litigation, *Lawrence E. Murr*

Vol. 21 What Every Engineer Should Know About Corrosion, *Philip Schweitzer*

Vol. 22 What Every Engineer Should Know About Lasers, *D. C. Winburn*

Vol. 23 What Every Engineer Should Know About Finite Element Analysis, *edited by John R. Brauer*

Vol. 24 What Every Engineer Should Know About Patents, Second Edition, *William G. Konold, Bruce Tittel, Donald F. Frei, and David S. Stallard*

Vol. 25 What Every Engineer Should Know About Electronic Communication Systems, *L. R. McKay*

Vol. 26 What Every Engineer Should Know About Quality Control, *Thomas Pyzdek*

Other volumes in preparation

WHAT EVERY ENGINEER SHOULD KNOW ABOUT PATENTS

SECOND EDITION

William G. Konold
Bruce Tittel
Donald F. Frei
David S. Stallard

Wood, Herron & Evans
Cincinnati, Ohio

MARCEL DEKKER, INC. New York and Basel

Library of Congress Cataloging-in-Publication Data

What every engineer should know about patents / William G. Konold . . .
[et al.] . –2nd ed.
p. cm. –(What every engineer should know ; vol. 24)
Includes index.
ISBN 0-8247-8010-8
1. Patent laws and legislation–United States. 2. Intellectual
property–United States. 3. Engineers–United States–Handbooks,
manuals, etc. I. Konold, William G. II. Series: What
every engineer should know ; v. 24.
KF3114.8.E54W48 1989
346.7304'86–dc19
[347.306486] 88-22654
CIP

MARCEL DEKKER, INC.
270 Madison Avenue, New York, New York 10016

Current printing (last digit):

10 9 8 7 6 5 4 3

PRINTED IN THE UNITED STATES OF AMERICA

Preface

This book provides a general outline of the law of intellectual property with particular emphasis on patent law. It has been written primarily for the engineer but will also be of general interest to others who need to be acquainted with the subject, including lawyers, mechanics and technicians, business persons, chemists, physicists, the simply curious, and of course inventors in general. The objective is to provide an appropriate perspective on patents, trademarks, trade secrets, and related matters, without undue use of specialized legal language and terminology.

To the best of our ability, we have accurately stated the law. The reader should be warned, however, that the law differs from court to court in the United States. Further, the law and practice change with new decisions from the courts, new laws enacted by Congress, and new rules of practice adopted by the U.S. Patent and Trademark Office. We therefore strongly recommend that when any question arises concerning a matter involving intellectual property, a patent lawyer be consulted. The patent lawyer, once given all of the specific facts pertaining to the matter, can advise the reader as to the best course of action for resolving the matter at hand.

It has been our intent, in this broad-spectrum discussion of patent law, to give the reader a sense of what patent law is all about. This general understanding should take much of the mystery out of patent law, enabling the reader to avoid pitfalls such as inadvertently dedicating intellectual property to the public or embarking on a course that will inevitably create a conflict with a competitor. In addition, general understanding should enable the reader to communicate more effectively with a patent lawyer when the need to communicate arises.

We hope that it will be understood that we do not intend this to be a definitive treatise on the patent law—or of intellectual property. Rather, we have intended to expose the reader, and particularly the engineer, to some of the areas of intellectual property law that may confront and concern him.

We have consistently used the male pronoun throughout the text for simplicity. Thus, we feel and hope that no slight will be felt by the female reader where we rely on our early Latin and English teachings of combined gender for the pronoun "he" as used here.

Concerning the revisions, our book was first published in 1979. Since that time there have been several changes in the patent and copyright law and procedures, foreign and domestic, to say nothing of the cost of obtaining protection. The revisions update the original text and introduce the reader to improved searching facilities (Chapter 5); post-issuance practice, including maintenance fees and reexamination procedure (Chapter 8); patenting biotechnology inventions (Chapter 10); Patent Cooperation Treaty practice for obtaining foreign patent protection (Chapter 16); and software protection, including semiconductor chip protection (Chapter 19).

William G. Konold
Bruce Tittel
Donald F. Frei
David S. Stallard

Contents

About the Authors

The authors are partners in the law firm of Wood, Herron & Evans, Cincinnati, Ohio, and limit their law practice to the field of intellectual property, particularly including patents, trademarks, copyrights, and trade secrets. Each is admitted to practice before the U.S. Patent and Trademark Office as well as the Supreme Court of Ohio and various federal courts. Collectively, the authors have undergraduate training in the areas of chemistry, physics, and electrical and mechanical engineering.

William G. Konold obtained a Bachelor of Electrical Engineering at Cornell University and received his Bachelor of Laws degree from George Washington University.

Bruce Tittel is a graduate of Vanderbilt University with a Bachelor of Arts in chemistry and a Bachelor of Laws degree.

Donald F. Frei obtained a Bachelor of Mechanical Engineering degree from Cornell University and a Bachelor of Laws degree from Georgetown University.

David S. Stallard has a Bachelor of Science degree in physics from Georgetown College and a Juris Doctor degree in law from the University of Louisville.

WHAT EVERY ENGINEER SHOULD KNOW ABOUT

PATENTS

1

Types of Intellectual Property

There are various types of intellectual property with which the engineer may be concerned. Principal among them are patents, which is the main focus of this book. But the engineer should have at least a smattering of knowledge concerning trade secrets, copyrights, and trademarks. Three brief chapters (18, 19 and 20) will treat these forms of intellectual property in greater detail.

This chapter, therefore, is to let you know what you are in for in the succeeding chapters, namely, a lot about patents and a little bit about trade secrets, copyrights, and trademarks.

For the moment, it will suffice to know:

A *trade secret* is information which a company, individual, or the like does not wish to have in a competitor's hands. It may include financial information, customer information, blueprints, and the like. In the final analysis, a trade secret is what a court says is a trade secret.

A *copyright* is a statutory right to exclude others from copying your creative works. Creative works which are protectable include writings, such as this book, works of art, three-dimensional works of art, photographs, compilations of information, and others which will be discussed in later chapters. When you publish your work, unless you place a copyright notice upon the work, you may put that work in the public domain so that it may be copied by anyone, with certain exceptions (see Chapter 17). The copyright notice may be in one of several forms, but if the work bears the word *Copyright* and/or the symbol © with the name of the copyright proprietor, and the year of publication, the federal copyright will be preserved.

A *trademark* is a mark, word, symbol, or device which is applied to goods moved in commerce. A trademark is, in a sense, a designation of the origin of the goods and, in many instances, is an extremely valuable piece of intellectual property. Consider for a moment the value of the trademarks Coca-Cola and Kodak. Common law rights to a trademark are obtained as soon as goods bearing the trademark are injected into the stream of commerce. After the mark is applied to goods moving in interstate commerce, it is possible to obtain a federal registration of the trademark, which of itself has considerable value.

But enough of these other forms of intellectual property–let us get on to our main topic, the patent.

2

What Is a Patent?

A *patent* is a grant by the United States government through the Patent and Trademark Office (hereafter U.S. Patent Office) to an inventor of the right to exclude others from making, using, or selling his patented invention for a period of 17 years (in all but design patents, for which the period is 14 years). The patent does *not* carry with it the right to manufacture the patented invention free from infringement. This sentence holds a concept which some people have had difficulty understanding, and so let us give you an example. Let us assume that the Ajax Stool Company has obtained a patent on a four-legged stool consisting of a seat and four legs. That structure will be considered to be the pioneer patent.

A couple of years later, the Acme Chair Company invents an improvement consisting of mounting a back on the four-legged stool. Acme Chair Company patents a chair comprising a seat, four legs supporting the seat, and a back projecting upward from the seat.

A patent granted to Acme Chair Company will give Acme the right to exclude others from making, using, and selling the chair with the back. Acme Chair Company, however, cannot manufacture, use, and sell the chair without infringing the patent of the Ajax Stool Company, for the Acme chair utilizes the patented four-legged stool structure earlier patented by Ajax Stool. Ajax Stool can manufacture its stool free from infringement because there is no earlier patent on the stool, but Ajax Stool cannot apply a back to the stool, for then they would infringe the patent of the Acme Chair Company.

Does that clear up the whole subject?

A patent is granted to the *original inventor(s)* (joint or sole) in the United States (not true for all countries). The applicant for a patent must be an inventor–a real person just like you. Except in special cases, such as death of an inventor, the actual inventor must sign formal papers making oath to or declaring his original inventorship. In practice, when a company takes out a patent, the inventor or joint inventors execute the required formal oath and also execute an assignment of the invention to the company that owns the patent rights. Except in cases where an inventor is legally obligated to his* employer company and for some reason refuses to sign such formal papers or cannot be located, a company cannot apply for a patent without having the inventor sign the papers.

Remember, the patent must be filed in the name of the *original inventor(s)*. Therefore, you cannot patent subject matter that was created by others who did not take the trouble to patent it.

There are two types of patents with which the engineer is concerned. The first is a *design patent*, which is directed solely to the ornamental appearance of an article of commerce. In some instances, an "ornamental appearance" can also be the subject of copyright protection as a copyrightable work of art. The term of a design patent is 14 years. Remember: ornamental appearance only. The desirable functional, i.e., utilitarian, features of an invention will not impart patentability to it–only the ornamental appearance (Appendix A).

The second type of patent is a *utility patent* (Appendix B), generally more difficult to obtain and usually more valuable than a design patent. The subject of a utility patent may be an article, a process such as a chemical

---

*Throughout this book, the authors have used the masculine pronoun "he" for the sake of simplicity, with the understanding that it should always be taken as including women as well as men.

process or a process of manufacturing, an electrical circuit, manufacturing apparatus, a new composition of matter, or the like. Utility patents are divided roughly into mechanical, chemical, and electrical patents.

A patent usually takes the form shown in Appendix B, consisting of one or more sheets of drawings and a specification concluding in one or more claims.

The specification begins with a preamble in which the patentee provides a general description of the invention. There, he may describe the state of technology to date, termed *prior art* (prior patents, publications, public uses, and the like), the problem of the prior art, the reason the invention came into being, and in a general way, the features of the invention that distinguish it over the prior art.

Then, preceded by a brief description of the figures of the drawing, the inventor provides a detailed description of the invention. This description should be of the best form known to the inventor and should be sufficient to enable a person skilled in the art or field to which the invention pertains to practice the invention, either by duplicating the process of the patent or by manufacturing the article or apparatus. If the patent is insufficient in this regard, the U.S. Patent Office may refuse to grant the patent or, alternatively, if it is granted the patent may later be held invalid by a court.

The description does not have to be in such meticulous detail that any untrained person can pick up the patent and put the invention to use. Rather the description assumes that the reader is skilled in the art and has a general knowledge of what has gone before. Similarly, the drawings are not like detailed blueprints. Rather, they are pictorial representations of the invention, many times using perspective views, appropriate shading, and the like to enable the reader to *easily* perceive what the invention is all about.

There are several types of persons who will read the patent and to whom the patent should be addressed. The patent application must be read by an Examiner in the U.S. Patent and Trademark Office, the Examiner being the person who makes the initial determination as to whether or not the patent should be granted. The description must state to the Examiner with clarity and completeness what the invention is all about. The patent is read by patent lawyers, generally with some training in engineering, and they too must understand the invention which is described.

In the final analysis, when the patent is asserted against an infringer, the patent may be read by a judge who probably has no engineering background, and he must gain an appreciation of the invention described in the patent. The judge will usually be assisted by the testimony of engineers and the argument of lawyers who represent the patentee (owner of the patent) in litigation. Nevertheless, if the subject matter is obscurely written, the judge may be inclined to hold the patent invalid. On the other hand, if the description of the invention and the advance for which it stands is written with clarity, the judge is more likely to hold the patent valid. It therefore behooves the engineer, as he reviews the writing of the lawyer preparing the patent application, to be sure that his invention is described fully, accurately, and with clarity.

The claims, which constitute the final paragraphs of the specification, consist of a succinct statement or description of the novel elements or steps by which the invention is distinguished from all the prior art that has gone before it. The claims define the parameters of the patent grant, i.e., what the patentee may exclude others from doing. Stated another way (and admittedly this is an oversimplification, but at least it is a good starting point), if it is possible to read the claim in its terms on a competitor's product, apparatus, or process (i.e., the claim's words describe it) the competitor is considered to be an infringer.

Therefore, in working with your patent lawyer in the preparing and obtaining of a patent, it is extremely important to be sure that the claims, or at least the broadest of them, do not contain undesirable limitations which can easily be omitted by the competitor, thereby enabling the competitor to avoid infringement. Most patents will contain more than one claim. If a competitor's product, apparatus, method, or composition of matter is covered by one claim and not covered by all of the remaining claims, the competitor still may be an infringer because each claim stands alone in the determination of the question of infringement.

It is important that all of the claims collectively cover all of the important features of the invention.

Only One Invention to a Patent

In many instances, a complex invention such as complicated apparatus will involve more than one invention. All inventions embodied in such apparatus will normally be described in a single patent application because all inventive components of the apparatus must be shown together in order to provide a complete description of useful apparatus. Where all components have been invented by a single entity (joint or sole inventors), a single patent application may have claims directed to all of the inventions.

Since the patent can have claims directed to only a single invention, the U.S. Patent Examiner will, in his first action, point out that there are multiple inventions claimed in the patent application and require restriction to one of them. In responding to the requirement for restriction, the patent lawyer will elect one invention. During subsequent prosecution of that patent application, only that one invention can be considered. However, it is possible to file one or more "divisional" applications, each divisional application being directed to one of the remaining inventions described in the original application.

Thus, a single application may spawn multiple divisional applications before the complete apparatus is satisfactorily protected by claims to all of the inventions which it incorporates.

Cost of a Patent

Three basic factors determine the cost of filing an application for a patent. The first is the government fee. If the owner is a small entity, a company not exceeding 500 employees, the fee is $170. If the owner is a large entity, exceeding 500 employees, the fee is $340. The second is the cost of drawings, which normally should be prepared by a draftsman skilled in making patent drawings in accordance with U.S. Patent Office rules. The third is the time spent by the patent lawyer preparing the patent application. This third charge is widely variable, depending upon the complexity of the subject matter, but is normally directly dependent upon the time the lawyer expends in the preparation of the patent application.

There are also prosecution expenses, which are dependent almost entirely upon the time the lawyer spends writing amendments (briefs) or conducting interviews to persuade the U.S. Patent Office Examiner to grant a patent. Finally, there is a government base issue fee of $560 or $280 depending on whether the owner is a large or small entity.

In general, the patentee may expect to spend no less than $1500 in the preparation and filing of a patent application, no less than $500 in prosecuting the patent application.

The numerical value for these fees is considered reasonably accurate as of the date of the revision of this book but fees will no doubt change as the cost of living varies.

Marking a Product "Patent Pending" or "Patent No. _______"

You will see many products bearing the words "patent pending," "patent applied for," or like designation. These designations have no legal effect other than to inform the purchaser that there has been a patent application filed at the U.S. Patent Office. The potential copyist should therefore have some concern that if he tools up for copying, a patent may issue and he will be enjoined from further manufacture.

The words "Patent No. 3,456,789" applied to a product should mean that a patent has been granted covering the product.

Placing the patent number on a product can be very important for, by statute, it puts the public on notice that the product is patented and any infringer is liable for damages from the date the infringer first puts the infringing product on the market.

If the product did not have the patent number on it, the infringer would be liable for damages only from the date he was given actual written notice of infringement.

Care should be exercised in marking products with the designation "patent pending" or with a patent number. A patent application must actually be pending when the designation is used and, similarly, a product marked with a patent number must actually be covered by the patent. Otherwise, the person putting the designation on the product is liable for fines for false marking.

3

What Can Be Patented?

From time to time we hear engineers say, "Oh, you cannot patent that." Don't be too sure! There are many categories of subject matter that can be patented. The U.S. Patent Office has in fact classified patentable inventions to about 400 general classes containing about 30,000 subclasses. Perhaps your invention falls within one of the 30,000.

Specifically, Section 101 of the patent law states that the following can be patented:

> *Inventions Patentable*
>
> Whoever invents or discovers any new and useful process, machine, manufacture, or composition of matter, or any new and useful improvement thereof, may obtain a patent therefor, subject to the conditions and requirements of this title.

After the patent laws tell you what can be patented, they then state what cannot be patented in Section 102 and 103. Section 102 has six categories of statutory bars to patenting. Rather than quote the law and leave you as bewildered as the patent lawyer is, let us suggest that it boils down to this:

If the subject matter was first invented in this country by someone else, you cannot obtain a valid patent on it. Further, if the subject matter was described in a printed publication anywhere in the world or was in public use or on sale in this country more than one year prior to the filing date of your patent application, you cannot obtain a valid patent on the subject matter. For all of the various ramifications and nuances attendant to Section 102, you had better see your patent lawyer.

In Section 103, the patent laws give you one more extremely important criterion for determining patentability. It is the simplest to understand and the most difficult to apply of all the criteria, i.e., a patent should not be granted "if the differences between the subject matter sought to be patented and the prior art are such that the subject matter as a whole would have been obvious at the time the invention was made to a person having ordinary skill in the art to which said subject matter pertains."

The key word is *obvious*.

Your coming to grips with this term may be the single most important thing you can understand or do in your association with the world of patents. For example, as you are attempting to assist your patent lawyer in the drafting of a patent application, if you can articulate for him what features of your invention would not have been obvious to persons of ordinary skill in the art–and why–you will have given him one of the most important tools in the drafting of the patent application.

During the prosecution of the patent application, when the Patent Examiner has cited prior art and the lawyer must distinguish over it, it may very well be up to you to tell your patent lawyer why your invention would have been unobvious to a person of ordinary skill in the art even though that person may have had before him the prior art cited by the Examiner.

From the foregoing, you now know in a general way what can and cannot be patented. There are some things which specifically cannot be patented.

Foremost among them is the perpetual motion machine. Most patent lawyers will, at some time, have to discourage an inventor who has come to the lawyer with a perpetual motion machine. A second category of things not patentable are methods of doing business. A third category is printing–pure printing. Strangely enough, the U.S. Patent and Trademark Office has a class of patented subject matter entitled "Printed Matter." However, the patents in this class relate to the use of printing where it is related in a novel and unobvious way to structure to which the printing is applied. Here again, we suggest that you see your patent lawyer to straighten this one out for you.

Another category of subject matter which may or may not be patentable is computer software. If the software is a pure algorithm, it is not patentable. If in some way the software, the algorithm, or what-not can be related to structure in a machine which processes the software, it may be patentable. It must, of course, meet the statutory requirements of Sections 102 and 103 referred to above. See your patent lawyer.

In the last several years the field of biotechnology has rapidly expanded. Until the 1980s, when the Supreme Court ruled to the contrary, the U.S. Patent and Trademark Office had taken the position that living, biological inventions could not be patented. Many patents on biological inventions have been issued since then on human and animal cell lines, on organisms, fused cell lines, plants, and processes for making them. One requirement is that they be laboratory-made and not found in nature. In order to patent a microorganism or cell culture, one must deposit such in an appropriate depository. The public may obtain access to this once the patent issues.

4

Documentation of Invention to Establish Your Rights Against All Others

There are two stages which you go through in the process of making an invention. The first stage, often referred to as *conception*, is the period when you are creating the invention in your mind, perhaps with the aid of pencil and paper. The invention may be the solution to a problem, a new product, machine, process, chemical composition, and so on. The second stage in the inventive process, which is known as *reduction to practice*, is the implementation of the invention, which you have conceived in your mind, by the physical act of constructing an operative prototype of your invention and demonstrating its usefulness for its intended purpose. Even a relatively crude physical implementation of the invention is sufficient to satisfy the reduction to practice requirement. It is not necessary that the equipment you build in the course of reducing the invention to practice represent an optimal design, e.g., is in a state in which it needs no further improvement and is ready for release to manufacturing. But in order to constitute a legal reduction to practice, it must be tested to demonstrate that it will work in the environment for which it is intended. Thus, the first phase of the inventive process—conception—is a mental act or vision, while the second stage—reduction to practice—is a physical act.

For certain purposes, the filing of a patent application consistutes what is known as a *constructive reduction to practice* and serves as a substitute for an actual reduction of the invention to a practical physical state.

To establish your early date of conception, which is required in several instances as explained below, you must proceed to reduce your invention to practice with *diligence*. Diligence involves substantially continuous activity on the part of the inventor or those working with him until the invention is reduced to actual or constructive practice. This activity from time of conception must in some way be documented so that it can be proved if necessary. The U.S. Patent Office and the courts are usually quite strict about diligence and have found diligence lacking where there were unexcused delays of as little as a week or 10 days. Of course, delays due to illness, component unavailability, and the like will usually be excused and will not operate to destroy diligence.

When Proof of Invention Date Is Necessary

It is often necessary in legal proceedings in the U.S. Patent Office, in the courts, or informally among individuals or corporations endeavoring to avoid a formal legal proceeding, to establish the date when an invention was made. This can involve providing the date of conception, the date when the invention was reduced to practice, and in some cases, whether the inventor exercised diligence in reducing the invention to practice. In many cases, the proof takes the form of suitable documents, testimony, and the like. Illustrations of instances when such proof is necessary, or at least desirable, are:

1. To antedate a publication or an earlier filed patent which describes, but does not claim, an invention you are seeking to protect in a pending patent application. This may enable you to show that you made your invention before the date of the publication or patent and thus obtain a patent on your invention notwithstanding the existence of the publication or patent disclosure describing it. This is done by filing in the U.S. Patent Office suitable proof of the date you made your invention.
2. To antedate the invention date of another applicant for a patent or patentee. This may arise when your patent application claims the

same invention as the other applicant or patentee and your proofs are needed to establish that you are the first inventor and are thus entitled to sole rights to the invention. This is done in what is called an *interference proceeding* conducted by the U.S. Patent Office in which at least one of the parties is an applicant for a patent and the other either an applicant or a patentee. It may also be done in a civil action in court when both parties are patentees.

3. To establish prior inventorship and avoid liability for infringing an issued patent of another when you are sued for patent infringement (or threatened with suit) by a patentee. This arises when his patent claims an invention which in fact was earlier completed by you and not abandoned, suppressed, or concealed. Incidentally, if you make an invention before another but abandon, suppress, or conceal it, you lose your right to obtain a patent, even though you are the first inventor, in favor of the subsequent inventor who obtained a patent and who did not abandon, suppress, or conceal his invention. In this way, the patent system promotes dissemination of technology via the issuance of patents and penalizes inventors who attempt to keep their inventions secret.
4. To establish prior knowledge by you of an invention which another party claims to have given you in confidence. Such proof may help you avoid liablity for breach of confidence and misappropriation of the third party's alleged trade secret or proprietary information.
5. To establish knowledge of an invention by you or one of your employees prior to the employment date of a new employee. This need may arise where the new employee was previously employed by a competitor and your competitor claims that the new employee conveyed the invention to you in violation of his duty to maintain certain inventions and proprietary information in confidence. Proof of your possession of the invention prior to employing a competitor's employee avoids liability for knowingly receiving the trade secrets of a competitor in violation of a confidential relationship which existed between the new employee and his former employer.
6. To determine correct inventorship, i.e., who really made the invention or who made it first, where two or more employees of the same company claim to have made an invention covered by a proposed patent application.

Written Proof Preferable to Oral Proof

Proof of conception, reduction to practice, and diligence can be in the form of the testimony, i.e., oral statements, of the inventor, but preferably others working with him who have knowledge of his inventive activities. However, it is preferable that the evidence be in tangible form, i.e., documentation, models, and so on. There is always the danger, when you are relying on the testimony of a witness to prove conception, reduction to practice, or diligence, that when you need the proof the witness will be unavailable or uncooperative due to death, illness, hostility, or disappearance. In addition, with the passage of time, memories fade and a witness's testimony becomes less credible. Written records, on the other hand, endure although care must be taken to protect them against loss or destruction as well as alteration.

Attributes of Credible Invention Records

1. Invention records should be sufficiently complete so that any other person skilled in the field to which the invention relates can understand exactly what was in the inventor's mind, in the case of conception, and exactly what physical acts were done in the case of reduction to practice. Where appropriate, invention records should emphasize critical aspects of the invention such as specific frequency, pressures, temperatures, voltage, power level and current ranges, and the like. Where possible, invention records should include blueprints, sketches, models, tabulations of experimental data, photographs of prototypes, laboratory set-ups, oscilloscope traces and the like, strip charts, computer printouts, etc.

2. Invention records, like any writing, should be definite, precise, clear, and organized. This permits others to completely and fully comprehend what the inventor invented.

3. Record should be made contemporaneously with the activities being recorded. In this way, important facts are likely not to be omitted. Contemporaneous records also tend to be more credible.

4. Record entries should be made in a bound book with numbered pages. This is in contrast to a looseleaf binder, which permits pages to be removed and is hence less credible. In making invention records, use a pen

and not a pencil which can be easily erased or altered. If an error is made, draw a line through it rather than erase it. When invention records have erasures in them it is often difficult to convince the U.S. Patent Office or a court that the erasure was made at the time the record was created rather than months or years later. A cardinal rule of invention record keeping is that you never alter a record at a later date. Also, do not leave blank pages or spaces since these can be argued to have been filled in at a later date.

Your invention record book should be kept intact by not removing pages. If blueprints, photographs, oscillograms, and the like originally recorded on another medium are incorporated into an invention record book, they should be securely fastened by taping or the like. In addition to securely fastening photographs, oscillograms, and so on in your book, you should sign and date each and refer to such items in your record notes. Record enough facts to justify any conclusions you make.

5. When making an entry in your record book, explain inconsistencies (real or apparent) vis-a-vis earlier record entries. This helps clarify the record.

6. Record what does not work as well as what does work. This is helpful in explaining delay in proving diligence as well as in proving that the invention is not "obvious" but rather was an invention made only with great difficulty and after a great deal of effort.

7. When in doubt, explain; and where possible, demonstrate to others. Then have your invention record entry read by others, preferably two persons who are capable of understanding it, and have them sign and date your entry at the end, noting that they have "read and understood" the invention. Those witnessing your record entries should *not* be persons who could conceivably be considered coinventors. Due to rules of the U.S. Patent Office, records witnessed by a coinventor are not admissible to prove conception, reduction to practice, or diligence. Such records have little, if any, real value in a U.S. Patent Office proceeding.

8. Where you witness conception or reduction to practice of an important invention of another, make reference to such fact in your own record book.

9. Record the problem or objective, its origination and your proposal, conceptualized solution, or invention. Also record your probable method

of implementation, other uses of the invention, and its advantages. Such is particularly useful in proving conception.

10. The following items, useful in proving reduction to practice, should be included in notebook entries:

a. Equipment used: its manufacturer, model, and photographs of set-up
b. Results: samples, models, oscillograms, strip chart records, and so on
c. Conclusions to be drawn, difficulties encountered, proposed changes or modifications
d. When and where acts were done
e. Names of those who did the actual work, and identification of those who authorized the work
f. Test data sufficient to show that the invention will work in the environment for which it is intended

11. To establish diligence, notebook entries should include all acts which show the progress of the invention between conception and reduction to practice. Useful entries include

a. Explanations of delays, if any, in reducing the invention to practice
b. Equipment ordered: what, when, how, and from whom; proposed use
c. Unsuccessful attempts to reduce the invention to practice
d. Work plans, time records

As alluded to earlier, because of peculiarities in the law dealing with proof of conception, reduction to practice, and diligence, oral or written evidence generated by the inventor (or a coinventor) requires *corroboration*, i.e., it must be supplemented, supported, and authenticated by independent evidence of a noninventor who is capable of understanding the technology involved. For this reason, all facts necessary to prove conception, reduction to practice, and diligence should be immediately reduced to writing or other documentary form and promptly signed by a noninventor capable of understanding the technology after he has reviewed and understands the invention record.

Factors to Be Considered in Determining Which Inventions Are to Be Recorded

1. The length of time the problem existed. Solutions to long unsolved problems are likely to be important.
2. Whether the invention is the only solution, significantly better than most solutions, or merely one of many equally good solutions. If the invention represents one of numerous solutions to a problem and is not significantly better than any of the other prior solutions, it is less likely to be patentable or of value to you than an invention which represents the only solution or is significantly better than others.
3. Savings in time and money provided by the invention. Obviously, the greater the economies provided by the invention the more likely it is to be patentable.
4. Importance to the company: prestige value of product; company needs a new product; invention has peculiar fit with company's existing product line or services; invention permits company to make entry into a field it wishes to penetrate.
5. Extent of use of invention in future: high or low volume; long or short term.
6. Whether the field of the invention is crowded or wide open. Often a relatively minor advance in a crowded field is patentable.
7. Likelihood parallel development programs exist in other companies. Such is often the case when a number of different companies are seeking to solve the same problem or are attempting to exploit a new technology.
8. Whether the solution is one which others skilled in the field are likely to develop in the normal course of their work. If so, the invention is probably not important or patentable.
9. Whether the invention constitutes a noninfringing design around a patent owned by a competitor.
10. The general need and usefulness of the invention.

Obviously, you cannot make elaborate and time-consuming records of every thought and act. If you did this, you would have little time for doing your day-to-day work and your productivity would be substantially reduced. You therefore must exercise good judgment in deciding what things you ought to record and how much should be recorded about them. Typically, engineers deplore making invention records and are reluctant to do so. It is often an area in which they have little experience and time. If you find this to be a problem, you might try dictating your thoughts with the aid of a tape recorder in a conversational mode and having them transcribed by a secretary. You can then fasten the typewritten record in your laboratory notebook and have it signed and witnessed in much the same manner as if you had written it by hand. You will be surprised how quick and effortless the invention record-keeping process can be if you do this.

5

Types of Patent Searches and What They Can Do for You

General Comments

Most patent searches have several aspects in common. With the exception of certain records data bases accessible by computer, each involves studying records and patents available in the U.S. Patent Office located in Arlington, Virginia just outside Washington, D.C. The U.S. Patent Office is the only place, at this writing and in this country, where the patents are classified by subject matter with all patents of each class and subclass being conveniently available for public searching.

Normally, the searches are conducted by a patent lawyer who travels to the U.S. Patent Office or by a lawyer's associate or professional searcher who resides there. Complete knowledge by the patent lawyer of the subject matter to which the search is directed is required, and the search cost will depend on a number of factors, including the type of search, the complexity of the subject matter, the adequate classification of patents in the areas of interest, and the number of patents which have issued in the areas of interest. Ordinarily, searches will result in a written opinion as to patentability of the subject searched, whether it can be commercialized free of infringement liability, or other features of the search in question. Copies of pertinent patents usually are enclosed with the search report. There are several typical types of searches which are discussed below.

Collection or State of the Art Searches

Let's suppose you work for a corporation whose management decides that corporate sales need a boost with a new product in a particular line. Or you work for a corporation having a multimillion dollar customer and your sales department has just promised that customer an improved machine product having a 15% increase in output in order to keep from losing the account to competition.

Responsibility for the new product or improvement has been assigned to you, a deadline is set, and all those midnight-to-dawn ideas somehow don't seem to be as promising in the light of day. Take heart, for you may find relief in a variety of ideas and suggestions derived from a collection or a state-of-the-art patent search.

A collection search is a form of patent history of a particular art. In a collection search, every patent relating to a particular apparatus, composition, process, or element of these is obtained. In sum, the collection of patents forms a developmental history of the subject matter in consideration and you may learn from this what others have considered worth patenting through the years or even what has not worked and has been abandoned. These patents may then be used as stepping stones toward the development of new products or improvement.

Available in the U.S. Patent Office are nearly 5 million patents, each classified into one of a variety of subclasses within some 400 subject matter classes. Basically, the patents are classified and cross referenced by the subject matter of their claims. More particularly, the patents are classified by apparatus, structure, composition, or method (process) to which the patent is directed. Still further, and depending on the subject matter classified, a searcher may be able to locate subclasses containing patents directed to an overall apparatus or process as well as subclasses directed to the pertinent component elements of an apparatus or a process. Thus it is usually possible to locate substantially all of the patents relative to a specific subject matter. The U.S. Patent Office classification of Artificial Body Members, Class 3, is relatively short but illustrative, and is attached as Appendix C.

Without getting into the actual mechanics of a search item, suffice it to say that in most cases it is possible to obtain at least several patents having pertinent disclosures relative to the general type of new product or apparatus in which an improvement is desired.

You should remember several important aspects of a collection search and the nature of its results. First, not every machine, apparatus, composition, or process of importance has been patented. While a collection search normally provides at least several pertinent patents, there may be some commercialized apparatus or process on which no patent application was ever filed. Yet, such nonpatented apparatus or process is part of the prior art.

Second, not every patent discloses something of significant commercial importance or usefulness, and there may be excess baggage in your search results. Some discretion will have been applied by the lawyer in selecting pertinent patents. Even further discretion will be exercised by you in extracting useful information from the patents located.

Third, a collection search may be extended beyond the U.S. patents available in the U.S. Patent Office's public search room to the Patent Examiner's files. These include a large number of foreign patents from a variety of countries (although a complete collection of all pertinent foreign patents from any particular country cannot be guaranteed).

Fourth, a collection search can be "customized" to cover either a broad or a limited area. For example, it is possible to collect all patents claiming rotary valves or only those patents relative to butterfly valves which are a specific type of rotary valve.

Finally, and perhaps most important, you should note that a collection search is usually most helpful at the *beginning* of a project. It can (1) prevent you from expensive (both time and cost) developmental plodding into previously well-plowed ground by showing you what others have tried with success or failure; (2) by its lack of information, suggest undeveloped areas in which further development or innovation may be useful; and (3) suggest the scope of protection available for further patent protection and at the same time warn you as to what patent coverage others have obtained so that development of infringing subject matter might be avoided.

Patentability Search

Now that you have relieved your management by providing the new or improved widget within the deadline, everyone wants to know if it can be patented. A patent search is a search conducted for the purpose of determining, first, whether or not the invention can be patented and, second, what the scope of that protection will be.

If a collection search has been conducted, it can be used as a basis for a patentability search and no further searching may be needed. More often than not, however, you have devised some ingenious feature far beyond the scope of ordinary or previous human capability and a new or updated search of the patents is necessary to provide a thorough basis on which a decision to file a patent may be made.

Now is the time to determine if the invention is a stroke of your exclusive genius or if some poor, unsophisticated soul just happened upon it in another time and place. Keep in mind that despite your intricate knowledge of the product area, not all patented items are commercialized. It is entirely possible that this search may develop very pertinent prior art which will eliminate or limit the likelihood of obtaining a worthwhile patent on your invention despite your independent development of it!

Once the patentability search is completed, the patent lawyer will report the results to you or to management, usually in writing and usually with an indication of the potential scope of coverage which the U.S. Patent Office would be justified in issuing. Having this report in mind and assuming some feature is potentially patentable, management can make a decision as to whether a patent application should be filed. As you will learn in a later chapter on trade secrets, sometimes an invention, even though patentable, is best protected as a trade secret and should not be made the subject of a patent.

Apart from providing patentability information to you, a thorough patentability search is a very valuable tool from two other aspects. If a patent application is to be filed, the results of the patentability search provide an invaluable aid to the patent lawyer in drafting the application. First, it helps him define the prior art and the background of your invention, enabling him to highlight the patentable features of your invention. Second, and assuming the most relevant prior patents are located in the search, the claims of the application can be drafted to avoid being too broad, which might make then invalid, and yet not so limited as to fail to provide the scope of protection deserved. Thus, a patentability search can be the first major step in obtaining a strong valid patent.

Your patentability search and its results will be much more accurate, thorough, and complete if you are able to communicate fully to the patent lawyer not only the details of what you have done but what the background of your invention is, including any known previous pertinent patents or disclosures and work by others. This will enable him to determine the nature and extent of the search required and the importance of the features you have invented. An objective evaluation of these factors will serve to highlight at an early time the potentially patentable differences between your invention and the existing state of the art, and will aid him in interpreting the patents he studied during the search.

For cost-conscious engineers, and we have yet to find a working one who isn't or won't be in the future, a patentability search usually costs something less than a patent application, and many times substantially less. It expense is increased (1) where it is important to be thoroughly informed on the peripheral and incidental patentability aspects of an invention; (2) where the prior patents to be searched are either inadequately classified and many subclasses have to be considered, or there is a large amount of patent activity in the area and many patents must be studied; or (3) where the subject matter is complex and lengthly patents must be read in detail.

On the other hand, and in most cases, the cost of a patentability search is justified when compared to the normally higher cost and risk of preparing and filing a patent application without knowing whether or not the invention is anticipated or rendered obvious to one of ordinary skill in the art (and thus unpatentable) by clear disclosures in prior patents. Even if the cost of a patentability search is high, it is in many cases justified when compared with the expense and effort of obtaining a patent or when a sound patent position in the field is to be obtained.

For these reasons, a patentability search is normally recommended. Only in those cases where the invention may have little commercial value and where the cost of filing a patent application is low will a patentability search perhaps not be justified.

Infringement Search

Now that your new widget, composition, or process is conceived (and perhaps engineered to some extent) and it looks as if it could be commercialized, you need to know whether you can make it, use it, or sell it without infringing the unexpired patents of others. Since the ordinary legal remedies sought for patent infringement include injunctions against further manufacture, use, or sale, and damages (see Chapter 13, "Enforcing Patent Rights"), the necessity of an infringement search is obvious. In those cases where initial manufacturing costs, such as for tooling, are high, where extensive inventory is necessary, or where large sales are anticipated or depended upon, ascertaining your freedom from infringement liability or accurately evaluating your risks is particularly important. Also, your invention may be the subject of patent licenses or agreements with others and it will be necessary to include warranties of noninfringement. A search may also be necessary to insure that the subject matter you are acquiring is not covered by other patents as well. An infringement search is an important part of a decision to manufacture, use, or sell an apparatus, composition, or process or to take license or assignment therefor.

As with the patentability search, any prior patent search is of value to the infringement search as a starting point. Except in an unusual case, however, the complete infringement search is generally more time consuming than the searches previously discussed. In order to determine whether infringement hazards exist, the patent lawyer must know the details of the apparatus, composition, or process in question, and he must study the claims of all pertinent patents to determine if they cover any of these details or the subject matter as a whole.

During the infringement search, the patent lawyer will first study the patents classified in the fields pertinent to the subject matter you wish to make, use, or sell. If pertinent patents are found, he will also study the patents cited by the Examiner during the examination and granting of them. If any patent contains claims broad enough to cover your subject matter, the patent lawyer will also normally study the U.S. Patent Office's file of that patent which was made during its prosecution (called a *file wrapper* or *file history*). He will be looking for a variety of things including, to name a few, whether any material in the file creates an *estoppel*, i.e., limits the patent's claims so they do not cover your subject matter, whether there are grounds rendering the application or the patent issuing from it defective or invalid, and whether the Examiner's search for prior patents included the most pertinent patent classes and subclasses.

Once the patent lawyer completes his search and investigation, he will normally render an opinion as to whether the subject matter in question can be made, used, or sold without incurring infringement liability. If no patent infringement hazards are located, the coast is clear. On the other hand, if hazards are located, his opinion will point them out and he may render advice on which you can assess the risks generated by the infringement hazard.

In some cases, the patent lawyer may be able to suggest alternate constructions or process steps which would avoid infringement. In others, he may indicate that one or more claims are infringed, but that they are, in his opinion, invalid for specific reasons. If he feels the claims may be invalid but he is not sure since the issue of validity is a close one, you may wish to redesign to minimize the risk of incurring infringement liability or you may continue on the program to commercialize the subject matter without redesign, taking the business risk that the patent owner may sue for infringement and you would have to defend on the basis that the claims were invalid. Such a decision should be adopted only upon the advice of a competent patent counsel who has thoroughly studied the matter.

In an infringement investigation, patents classified in pertinent subclasses are normally studied in more depth than they are in the previously discussed searches. As a result, you should acquire a sound legal opinion as to your freedom to make, use, and sell the questioned subject matter or as to your potential infringement liability.

Validity Search

You now have been advised that your widget, promising as it may be, infringes a patent owned by another party. Management is unhappy and your development efforts appear to have been wasted. Redesign is either too expensive or the patent is so broad that even redesign will not avoid infringement.

Perhaps you need to know whether a broad patent covering your competitor's hot-selling item is valid so as to give him a justifiable right to enforce it against you should you duplicate his product. Or perhaps you have entered the market with your product and are served with a complaint for patent infringement.

It's not yet time to throw in the T-square or Bunsen burner since you have not yet determined whether the patent is valid. In each of these cases, a validity search may establish that the patent is invalid.

In this search, the patent lawyer will have the specific problem patent in mind and his efforts will be directed toward locating prior art patents or other facts which will anticipate the claimed subject matter of the problem patent or which will render it obvious to one of ordinary skill in the art, either of which would invalidate the patent (see Chapter 3, "What Can Be Patented"). He will normally start by studying the problem patent, studying its file in the U.S. Patent Office, and studying the patents cited by the Examiner during his consideration of the application from which the problem patent issued. In addition to his search for pertinent prior art patents, he will also consider whether the specification and claims are appropriate according to the patent laws.

Many items may provide the basis for a validity opinion in addition to patents. For example, if you know the claimed subject matter was published or sold more than one year prior to the patent's filing date, that fact should be brought to the patent lawyer's attention. Accordingly, old sales catalogs, trade journals, prior literature, and your general knowledge of the field can be very valuable in determining whether a particular patent claim is valid.

Once the validity search is complete, the patent lawyer will prepare his report. Even if the patent claims are clearly invalid, the patent is presumed by law to be valid, and infringement is accompanied by the risk of defending a patent suit. An opinion of invalidity, based on thorough research and soundly supported, however, may justify the risk of having to defend an infringement suit or may bring to light facts which would render any enforcement attempt by the patentee so unjustifiable as to give you a basis for counterclaim if you were sued.

Accordingly, a validity search may provide an absolute defense to a potential or existing infringement suit based on a particular patent. It offers an opportunity to minimize the difficulties raised by the problem patent of another by providing information which may render it invalid.

Foreign Patent Searches

If for some reason, which occasionally arises, a comprehensive search of foreign patents is necessary, such can be obtained by your patent lawyer through various associates and search bureaus to which he has access in foreign countries.

Inventor and Assignee Searches

In some cases it is desirable to determine whether or not a particular inventor patented certain subject matter, or whether a company or person owns a patent directed to specific subject matter. A specialized search can be conducted to

1. Locate all patents in which a certain known person was named as a sole or joint inventor
2. Locate all patents presently owned by a company, organization, or person

These searches are thus specialized tools for locating patents when the inventor or owner is known. In addition, these searches may be used to complete any of the other types of searches when it is known that a particular inventor or company is active in a particular field.

Use of Automated Data Bases and Related Services in Searching

Over the last several years, a number of companies have initiated the keying of patents into a number of different data bases which can be addressed by an appropriate terminal and phone line. These are available to public access upon payment of user fees, or through companies providing searches of their own data bases for a charge. Moreover, patents in several technical areas have been produced in machine readable form for use in the Patent and Trademark Office.

An automated search into one or more of the available data bases can be very helpful and can save a great deal of time. On the other hand, there are many types of searches for which the automated data bases are not particularly helpful. For example, if patents or technical publications which you are interested in locating are likely to contain specific key words, then a search of the data bases will easily locate patents or other documents having these key words. On the other hand, if not properly limited, such a search may generate too many listings for practical review. Automated searches then, as a rule, may be much more helpful in the chemical arts, where specific chemical names can be used as a search "inquiry," than in the mechanical arts, where the subject matter sought does not frequently lend itself to specific key words or modifiers.

Moreover, while the companies that offer data bases for use are continually adding patents to them, most are not currently so extensive as to include full-text patent searching for all unexpired patents, nor expired patents. In a few years, selected data bases may have all unexpired patents due to the constant addition of all newly issued patents and the passage of time. The ability to search all issued patents, expired or unexpired, in any technical area, however, is not now available.

Automated searches are extremely helpful in many specific areas which before were not possible. For example, the data bases now make it easy, at least for patents issued over the last few years, to generate listings of all recent patents issued to certain inventors, corresponding patents filed in foreign countries, issued patents citing another patent, classes of classification for further searching, and other informational requests of similar nature, which are handled so well by the automated searching of electronic data bases. Moreover, automated searches can be very helpful in most technical areas as a lead-in to conducting a typical manual search or as a follow-up to locate patents or references not located by a manual search.

Several companies offering extensive patent data bases are Derwent, Inc., Pergamon International Information Corporation, Mead Data Control, etc. Many patent attorneys maintain computer terminals and other equipment capable of addressing the data bases of these or other companies for the purpose of providing complete search services as may be required.

In summary, several distinctions and comments can be made. A collection or state-of-the-art search is helpful as a finder's tool, particularly at the beginning of a project. A patentability search is normally recommended, once an idea has been conceived, to determine the scope of available patent protection and to aid in the preparation of a patent application. The need for an infringement search, conducted to determine potential infringement liability, is not negated by a patentability search, which is conducted only to determine the likelihood of obtaining protection. Remember, patentability of your invention is not a bar to a suit against you for infringement of another's patent broad enough to cover your invention. A validity search is conducted to determine the validity of a particular patent you are concerned about infringing, and inventor and assignee searches are specialized searches conducted to locate specific patents of an inventor or patent owner. Each search is a potentially valuable and, in many cases, a necessary tool in determining patent rights and liabilities.

6

Preparing the Patent Application

Preparation of a sound patent application requires a high degree of skill and technical expertise and is a serious undertaking in which you should be able to make a valuable contribution.

While the content of the application is important in satisfying the requirements of the pertinent laws and rules of the U.S. Patent Office, an application should always be drafted with its enforcement in mind and with a view toward strengthening it against challenges of ineffectiveness or invalidity which are to be expected if it is brought to suit.

Basically, the application contains three major parts:

1. A specification
2. A drawing, if the nature of the invention admits of a drawing
3. An oath or declaration by the inventor(s) containing certain legally prescribed statements.

The Specification

Our present patent laws provide in Section 112 that:

> The specification shall contain a written description of the invention, and of the matter and process of making and using it, in such full, clear, concise, and exact terms as to enable any person skilled in the art to which it pertains, or with which it is most nearly connected, to make and use the same, and shall set forth the best mode contemplated by the inventor of carrying out his invention.
>
> The specification shall conclude with one or more claims particularly pointing out and distinctly claiming the subject matter which the applicant regards as his invention . . .

It is accordingly important for the patent lawyer to understand the correct terms of the art or field in which the invention lies, as well as the best known mode for carrying out the invention, so that he is able to satisfy these requirements. Your aid in this regard will be very helpful in producing a well written application.

It is also important for the specification to tell a story, i.e., the specification should highlight the invention by describing its background and the problems to which the invention provides a solution. A description of the best known prior art and its deficiencies serves to aid in distinguishing the invention from what has been done before. Your knowledge of the invention's advantages, in contrast to the deficiencies of the prior art, should be communicated to the patent lawyer at the beginning to aid him in drafting the application.

Apart from the statutory and other prescribed formalities concerning the application, perhaps the singularly most important aspect of an application is its claims. It is the claims which define the scope of patent coverage, and it will be the claims which determine whether your patent covers the product of your competitor. The most desirable claim is one which provides the broadest protection deserved yet is not so broad as to be invalid because of prior art anticipation or suggestion. This is, in many cases, a very fine line requiring the utmost skill to develop. Your close communication with the patent lawyer preparing the application will be of great value in educating him as to the scope of protection your invention deserves and in drafting the correct claim language in a form which is neither too broad nor too narrow.

The importance of preparing a complete and accurate application for filing cannot be overestimated. Once the application is on file, you are normally stuck with it; later additions of new matter are not permitted.

The Drawings

An application usually contains one or more drawings when the subject matter is such that it can be shown in a drawing. Thus, a drawing is usually supplied in a mechanical or electrical apparatus application but is usually not supplied in a chemical composition application. A drawing comprising a flow chart, for example, might be used in an application directed to a process or method.

Generally, the patent lawyer will direct the preparation of the drawings by a patent draftsman familiar with the drawing formality requirements of the U.S. Patent Office. The drawing is not, in the usual sense, an engineering drawing; rather, its purpose is to illustrate the patentable features of the invention. Dimensions are not included, and breakaways, cross sections, perspectives, and exploded views are frequently used for clarity of illustration. Occasionally, for illustration, the drawings are not to scale, there being no scale requirement. In most cases, the elements of each figure of the drawings are identified with numbers and are discussed or described in the specification following a brief description of what each figure shows.

Since illustration is its primary function, the patent drawing is frequently shaded, general environmental structures are shown, and other niceties added. The chief engineer of a new client once had high praise for the drawing in a patent application he reviewed but remarked that the patent draftsman could not, with that type of work, hold an engineering job in his company where time and cost were of the essence. Stick to your engineering skills and let the patent draftsman use your engineering prints as a basis for his patent drawing.

The Oath or Declaration

Each patent application must usually include an oath or declaration, signed by the inventor(s), attesting to a variety of items. The inventor must state that he is the first inventor of the subject matter for which he seeks a patent and he must state, in particular form, a denial of knowledge of specific items which would render the application or patent issuing thereon improper or invalid. He must also identify any corresponding application filed in foreign countries. He must also acknowledge, in the oath or declaration, a duty to disclose information he is aware of which is material to the examination of the application.

Signing of the oath or declaration must take place after the inventor has read the application. Once the signing takes place, the application is not to be altered or partly filled in. Also, the application is not to be executed in blank. If any alteration or fill-in takes place after the signature to the oath or declaration, or if an oath is signed in blank, then a new oath or declaration should be executed after all corrections have been made and before the application is filed. It is now possible to file an application before execution of the oath or declaration by the inventor. Once the application is filed, a set time is granted to submit a signed oath or declaration. No changes or additions constituting new matter can be added. Failure to satisfy these regulations can result in the striking of the application from the files of the U.S. Patent Office or in the invalidity of any patent issuing on the application.

Reviewing, Revising, Signing, and Assigning the Application

Now that you have told your patent lawyer all you know about your invention and he has completed the application and has returned it to you for study and execution, you are most likely in for a pleasant surprise. It is not unusual for the engineer, particularly one with little patent experience, to be very pleased with or surprised at the clarity, simplicity, and thoroughness with which you now find your sophisticated and complex invention described and claimed. Don't forget, however, that the patent application is no place for unintelligible, fine-print language. Its prime functions are to (1) describe the invention in such full, clear, concise, and exact terms as to enable any person skilled in the art to which the invention pertains to make and use it and (2) to particularly point out and distinctly claim the subject matter you regard as your invention. The application is written for reading by engineers in the field and must be in language they understand. Now is your time to consider what the patent lawyer has done and to let him know of all your additions, deletions, or suggestions. If there is anything which is incorrect, incomplete, or which you don't understand, have it changed. The application is not written in granite, and no pride of authorship by patent counsel should stand in the way of changes necessary to complete the application or to render it accurate.

Usually, the application delivered to you will be in a format such as follows (not necessarily with the following subtitles):

1. Title of the invention
2. Abstract of the disclosure
3. Cross-references to related applications (if any)
4. Background of the invention
   a. Field of the invention
   b. Description of the prior art
5. Summary of the invention
6. Brief description of the drawing
7. Description of the preferred embodiment of the invention (and examples if a chemical invention)
8. Claims
9. Drawings (formal, inked, bristol board sheets)
10. Oath or declaration (for your signature)
11. Assignment form (if the application is to be assigned to your corporate employer, for example)

Only after you have carefully considered the complete application and have determined that it accurately and completely describes and claims your invention should you execute the oath or declaration form.

Should you have any deletions, additions, or corrections in any part of the application, check with your patent lawyer and make these before executing the oath or declaration form. Read this form carefully so that you know what you are signing. If it is in oath form, it will have to be notarized; if in declaration form, your signature, without notarization, is sufficient. In either case, sign your full first name, middle initial, and last name.

If you are employed by a corporation or company, you may have a written employment agreement to assign your inventions to your employer. An assignment form is usually provided for your signature at this time, and notarization is usually required.

A Few Words About Naming of the Inventor

Under our present laws, the actual single inventor or the actual joint inventors must apply for the patent. The determination of inventorship is a legal matter and is not a matter of emotion or of corporate politics. Nor, without more, can the status of inventorship be appropriately given to a person for the purpose of "recognizing" him, or for the purpose of gaining legal rights or avoiding legal impediments to the application.

Essentially, the inventor or inventors are those who actually conceived the subject matter of the workable invention to such a level that no further inventiveness is required to make or work the invention. In some cases there are single inventors, and in others there are two or more joint inventors, each of whom contributed to the inventive process resulting in the invention.

Routine efforts in drawing, constructing, or working the invention, once it has been fully conceived, are not a sound basis for according inventorship status to anyone not involved in the conception process. For example, if you conceive a new staple remover and direct your draftsman or your shop worker to draw it or to make a model, you are the sole inventor. The draftsman or shop worker, who exercises only routine skills in following your instructions, does not become a joint inventor. One becomes a joint inventor only if he makes an *inventive* contribution to some feature of the staple remover which is incorporated into the invention. By the same token, your superior who asked you to devise a new staple remover is not an inventor unless he contributed something more, as by suggesting a specific novel and unobvious feature.

Present patent laws permit correction for misjoinder of inventors, i.e., adding or deleting an inventor if a mistake is made. This is permitted only under specific circumstances, and an appropriate inventorship determination should be made at the outset with the aid of your patent lawyer. Under no circumstances can a patent application filed in the name of a single inventor be changed such that the originally named single inventor is removed and another substituted in his place. Also, it is not now necessary that every joint inventor named in an application be an actual inventor of every patent claim, as was the case in prior years.

Filing Fees

Finally, it is useful to note that not every applicant pays the same government application fees. Under current rules, a large entity filing a patent application will pay a larger fee than a small entity. If the application is assigned to a large corporation of over 500 employees, or if the inventor is obligated to assign the application to such a corporation, then the higher fee is to be paid. Others can file a small entity statement and pay smaller fees.

7

The Patent Application in the U.S. Patent and Trademark Office

Your application has now been on file for several months. The U.S. Patent Office, which annually receives about 100,000 applications, inspects the application for completeness, assigns it a serial number and filing date, preliminarily classifies it and assigns it to a group of examiners knowledgeable in the field to which the invention relates. One or more examiners examine the application to determine patentability of the invention described in the claims of the application.

The U.S. Patent Office: Initial Handling

But let's start at the beginning. The U.S. Patent Office currently employs hundreds of Patent Examiners (normally with engineering or scientific backgrounds, and possibly legal training), with clerical supporting staffs and numerous incidental service branches. The Examiners and their supporting staffs are divided into several examining groups within three broad categories: chemical, electrical, and mechanical. Each of the groups within each category is further divided into "art units" composed of several Examiners familiar with the patents classified in a particular art. Depending on the size of the art unit and on the number of patents within a class, an art unit may be assigned one class, several classes, or a portion of several classes.

When your application reaches the U.S. Patent Office, it is initially classified and that classification determines the art unit to which yours and all other applications directed to a similar art or field are assigned. Within each examining group, the Examiner's search files contain the U.S. patents and certain foreign patents and publications classified in the art in which he works. As a result of his experience in searching patents and in considering the applications of others in the field, the Examiner usually has some familiarity with the general field to which your invention pertains.

The Examiner's Search and Action

Assuming that your application has been properly classified and assigned to the appropriate art unit and the Examiner to whom it has been assigned has determined that your claims are all directed toward a single invention, he proceeds with his examination to determine if a patent should issue or not. In order to pass your application for issuance as a patent, he must make a determination that your claimed invention is subject matter which the law permits to be patented, is useful, is new, and is unobvious to one of ordinary skill in the art to which the invention pertains.

The Examiner familiarizes himself with your application and then proceeds to conduct his own patentability search, searching for patents or other prior art which might anticipate your *claimed* invention or which might suggest the obviousness of your *claimed* invention. When the Examiner completes his search, he determines which of your claims are in condition to be allowed and which are not, and he issues an "office action" or letter specifying his reasons for rejecting one or more of your *claims*. More often than not, you may expect one or more of your *claims* to be initially rejected over prior art, with the Examiner taking the position that the rejected claim is anticipated by a cited patent or rendered obvious by one or more patents.

The office action is ordinarily mailed to your patent lawyer and you are given a period of three months in which to respond, which is extendible for up to three additional months upon payment of an increasing fee for each month late. It is at this point where active "prosecution" of your application begins.

Response to Examiner's Action

An amendment prepared by the patent lawyer is the typical response to an office action. The amendment may make changes in the specification, claims, or drawings, providing no new matter is added. In many cases it will include a claim change in order to more particularly define the claimed invention, or to broaden it, depending upon the circumstances.

But what happened to that beautiful, carefully prepared, air-tight application? Certainly it's easy for the Examiner to read your application and then in hindsight find components of the invention in obscure prior art patents which would have led the ordinary engineer in your field in an entirely different direction than that you pursued. You might think, "What does a Patent Examiner know anyway, cooped up in the U.S. Patent Office and never having experienced the day-to-day frustrations of a research and development department working in my sophisticated and complex art?!"

Before you decide that all Patent Examiners are crazy too (everyone knows you inventors are a little "odd"), consider for a moment the fact that the Examiner did not reject your disclosure or your invention—he merely rejected your *claimed* invention. It is his job not to determine whether your invention is patentable but to determine whether your *claimed* invention is patentable. So be charitable, give the Examiner the benefit of a doubt, and reconsider the *claims* in light of his remarks to determine if the claim language could be broadly construed to give rise to his rejection. The Examiner has imparted the broadest possible meaning to your claim language; what can you do to amend it, if necessary, to avoid his rejection?

Your patent lawyer will work carefully with you on the response to the office action, determining the Examiner's position, evaluating the claim language, and perhaps proposing changes to place the claims in condition for allowance or arguments to show the Examiner why his rejection was in error. Keep in mind that no new matter can now be added to the application, however, and that additions or changes in the claims must be supported by disclosure originally filed in your specification.

If the Examiner has found prior patents disclosing subject matter which is very pertinent to your claims, it may be necessary to limit your claims to the more detailed features of your invention or to simply narrow the overly broad terms used in the claim. If the Examiner's application of the teachings of the prior art to your claimed invention is not particularly apt (yes, there are these occasions), it is necessary to prepare arguments showing the Examiner the error of his ways. Such arguments may be presented without a change in the rejected claims or in appropriate situations with broadened claims.

Rejections based on "obviousness" are sometimes difficult to overcome. The Examiner has decided that one or more prior patents suggest your claimed invention and you do not agree. Depending on the circumstances, your patent lawyer may decide to support his arguments with appropriate affidavits relative to the deficiencies of the prior patents cited, to the unexpected results attained by your invention, to the commercial success which your invention has achieved, to the long-felt need in the art for the solution your invention provides, to the tribute paid to your invention by others, or to any number of other factors which may support your position of the nonobviousness of your claimed invention to one of ordinary skill in the art.

These amendments, arguments, affidavits, and the like are generally submitted in writing for the Examiner's further consideration. If he maintains an objection, he will most likely issue a "final rejection" after which you must put the application in condition for allowance or appeal. Otherwise, the application is abandoned. Although amendments may be made after the final rejection, you have no absolute right to make amendments (as you did in response to the first office action) and your amendment may not be entered if it does not place the application in clear condition for allowance or in better condition for an appeal. Entry of such amendments is generally at the discretion of the Examiner.

Interviewing the Examiner

An effective way to prosecute an application after a rejection is to interview the Examiner, discussing his position in order to understand it accurately, and presenting your amendment, arguments, or facts for his consideration. Frequently, the patent lawyer will plan the interview with your aid and then conduct it on his own. At other times he may desire your participation, or that of others, for the purpose of submitting information to the Examiner.

At the interview, the Examiner may commit himself to allow your claims, he may maintain his position, or he may indicate that he requires further time for reconsideration or for more searching in light of the interview. In most cases, he will desire that a written amendment be filed (all transactions with the U.S. Patent Office are to be in writing) and in any event reference should always be made in writing to the substance of the interview.

Appeals

If the Examiner maintains his rejection and you are unable to secure allowance of acceptable claims, you must appeal to the Board of Patent Appeals and Interferences which will decide the case on the written record, briefs, and optionally, a hearing. Your written prosecution record must be complete at this point since the Board will not consider subject matter not presented to the Examiner.

If the Board affirms the Examiner's rejection, you have the right to further appeal to the U.S. Court of Appeals for the Federal Circuit. This appeal is also limited to the U.S. Patent Office's record in the consideration. Alternately, you may appeal to the U.S. District Court in the District of Columbia where you have a "new trial" and may in a proper case submit information not considered by the U.S. Patent Office. In recent years, this so-called new trial has become so limited that virtually all appeals go to the other court noted above.

Issuance of the Patent

On the other hand, if your prosecution efforts were successful and an appeal is unnecessary (or if your appeal was successful), your application is approved or allowed for issuance as a patent. Formal notices are sent, final fees are paid, and the patent issues.

A single, formally ribboned patent grant document is issued and copies are thereafter available to the public. Copies of your patent are placed in the Examiner's search files and in the public search files in the classes and subclasses in which it is classified and cross-referenced, and your patent adds to the body of patent information available for consideration by the public.

Up until the date of issuance, your patent application is maintained in secrecy by the U.S. Patent Office. Upon issuance, however, it becomes publicly available. If you have in the interim decided to maintain your invention as a trade secret, your application should be abandoned before the final fee is paid in order to insure that it will not issue and be made public (see the chapter on Trade Secrets). If a patent is issued, however, your exclusive right, of 17 years duration, to keep others from making, using or selling your claimed invention runs from the issue date. Also, as of this date you are entitled to mark on your invention that it is covered by a specific U.S. patent.

Improvements in the Invention After the Application Is Filed: Continuation-in-Part Applications

What if you develop significant improvements or changes in your invention after your application was filed and before it issues as a patent and you desire to cover the improvements with a patent? The new improvements cannot be added to the pending application because they would constitute new matter. They can, however, be incorporated into a separate, new, continuation-in-part application. This new application is essentially the same application as was originally filed but is supplemented to take into account the new subject matter.

Prosecution of the original application can be continued, if desired, or the original application can be abandoned in favor of the new continuation-in-part application. In either case, the new continuation-in-part application must be filed while the original application is pending. Thus, it must be filed before the original application issues as a patent or before its abandonment. A new oath or declaration is required in the new continuation-in-part application to cover the new subject matter.

A Word About Divisional Applications

The patent laws require that a patent cover only a single invention. Thus, if your patent application contains claims covering more than one invention, the Patent Office will require you to limit your application to one invention by issuing a "restriction requirement." For example, if your claims are drawn to cover both a product and a process for making that product, such as a process for making soap and to the soap composition itself, the Patent Office may require you to restrict your present application to either process claims or to product claims. If your application is restricted, you may file a divisional application on the claims not maintained in the original application. The claims of the original applications are divided—hence, the term *divisional.*

Like the continuation-in-part application, the divisional application must be filed during the pendency of the original application. Unlike the continuation-in-part application, the divisional application is an exact copy of the original application, differing only in the nature of claims to be prosecuted, and no new oath is required. Various legal aspects of these types of applications are highly technical and you should not form conclusions or take action regarding either without legal advice.

The Duty of Disclosure

During prosecution of the patent application, the applicant, the attorney or agent who prepares or prosecutes the application, and every other individual who is substantially involved in the preparation or prosecution of the application and who is associated with the inventor, with the assignee or with anyone to whom there is an obligation to assign the application, owes a duty of candor and good faith to the U.S. Patent Office. Currently, the U.S. Patent Office defines this as being a duty to disclose to the U.S. Patent Office information which is material to examination of the application. Information is material where there is a substantial likelihood that a reasonable Patent Examiner would consider it important in deciding whether to allow the application to issue as a patent.

It is mandatory that you (as the inventor, for example) make sure that all activities and materials relating to the invention be discussed with patent counsel with a view toward disclosure to the Patent and Trademark Office. The U.S. Patent and Trademark Office Patent Rules go to extreme lengths to make it clear that no patent will be granted on any application where there was fraud on the Office or the duty of disclosure was violated either through bad faith or gross negligence. It is extremely important that all matters to be disclosed are disclosed in writing to the Office, together with copies of all prior patents, publications, or other relevant documents so that the record reflects that all such known information was in fact disclosed.

In determining whether or not certain information should be disclosed, the skilled patent attorney must consider its pertinence to the application. Information not material need not be disclosed. It is recommended, however, that all information of which you as inventor are aware, should be disclosed to patent counsel for purposes of making the decision of disclosure.

It is important to note that even the unintentional failure to disclose material information might result in a decision to strike the application or to hold the patent invalid.

For example, if the information is highly material, the Office or a court may not need to find much substantive intent to withhold the information before reaching a decision that the failure to disclose the information renders the application stricken, or the patent invalid. On the other hand, if the information is of questionable materiality, but the Office or court finds an intent to conceal or withhold it, then the application can also be stricken or the patent declared invalid.

In summary, if there is any doubt, disclose the information. In any event, seek counsel on all information as to which there is any chance of a decision that it may be material.

8

After the Patent Issues: Maintenance, Correction, Reissue and Reexamination

The grant of a patent by the U.S. Patent and Trademark Office does not end the patent owner's effort to secure the exclusive right to the claimed invention during the patent term. Under rules established in the early 1980s, maintenance fees must be paid periodically to keep the patent in force. Moreover, other post-issuance procedures are available to the patent owner to correct defects in the patent. If there is a printing error, for example, the patent owner may correct the error by filing a Certificate of Correction which is then attached to the original Letters Patent and all further patent copies, once it has been approved.

Where the defect is believed to render the patent wholly or partly inoperative or invalid, then the patent owner may be able to file a reissue patent application, seeking correction.

Finally, and under other new rules established in the 1980s the patent owner, or any other person, may file a request for reexamination of the patent based on substantial new questions of patentability which were not of record in the patent file before the patent issued.

Maintenance Fees

Through the late 1970s, issued patents on which the full issue fee was paid remained in force for their entire term without the payment of any further fees. In the case of a "normal" or utility patent, that term was for 17 years.

Under the statutes enacted in the early 1980s, however, a patent owner must pay a prescribed maintenance fee three times during the life of the patent to maintain it in force. These maintenance fees apply to all patents based on applications filed on or after December 12, 1980 (with the exception of all design patents, plant patents filed on or after August 27, 1982, and reissue patents where the patent being reissued did not require any maintenance fee).

Generally speaking, the three maintenance fees must be paid to maintain the patents in force beyond 4, 8 and 12 years after the date of the patent grant. The fees themselves must be paid six months before these dates to avoid a surcharge. They can be paid within the six months immediately preceding these dates by payment of a surcharge.

The amount of the first maintenance fee ranges from $225 to $450; the second maintenance fee from $445 to $890; and the third maintenance fee from $670 to $1340, depending on the size of the business of the patent owner (i.e., whether it is a small entity, not exceeding 500 persons, or a large entity, employing more than 500 persons).

Of course, these fees are expected to change periodically, and have already increased about 10% between 1982 and 1985.

The specific amounts due, and the dates on which they are due, are the subject of detailed and changing Rules of the U.S. Patent and Trademark Office. The owner of any patent based on an application filed on or after December 12, 1980 should consult a qualified attorney respecting continuing maintenance of the patent.

Certificates of Correction

Occasionally, after issuance an error may be noticed in the final printed patent copy. Where the error arose from a mistake on the part of the

U.S. Patent and Trademark Office, it may be corrected without expense to the patent owner. Such errors might typically involve a printer's mistake such as, for example, omission of a word or a phrase which appeared in the original patent description or claims as submitted or amended.

Where the error is not the fault of the U.S. Patent and Trademark Office, but rather is the applicant's mistake, the Office may still issue a certificate of correction, upon payment of a fee. In such a case, the correction cannot constitute new matter or require reexamination of the application.

Finally, a certificate of correction can be obtained where the correct inventor or inventors were not named originally. This requires a showing that the original inventorship designation was erroneous and there was no deceptive intention on the part of the actual inventor or inventors. Correction can be made where an actual inventor was omitted, or where a person named was not actually an inventor, assuming of course in each case the error was without deceptive intention.

Reissue and Reexamination

Once the patent has issued, more substantive issues than those mentioned above and affecting the patent may be correctable or addressed in the U.S. Patent and Trademark Office by the patentee, or in some cases, by either the patentee or a third party. Two different and separate procedures, known as *reissue* and *reexamination*, provide varying avenues for obtaining U.S. Patent and Trademark Office consideration of certain matters. Each procedure is based on specific statutory showings which the requesting party must make, and each procedure has its own unique set of advantages and disadvantages for the patentee or party involved.

The particular application of these proceedings and their potential result is a legal matter requiring the attention and advice of skilled counsel. An in-depth legal discussion of these proceedings is thus not proper for this book. Nevertheless, a few comments of interest about each proceeding may be found informative.

Reissue

Under our present Patent Laws an original patent can be reissued for the same invention in the patent where the patent is, through error without deceptive intention, deemed wholly or partly inoperative or invalid by reason of a defective specification or drawing or by reason of the patentee having claimed more or less than he had a right to claim in the patent. For example, if a patentee discovers that his issued patent claims his invention too narrowly, or too broadly, as to render its validity questionable, or that there is some error in his original description or drawing, he can file an application to reissue the patent in correct form.

Several major considerations exist with reference to reissue proceedings. First, a reissue application can only be filed by the patentee. Another party can protest the grant of a reissue, but cannot initiate or request reissue of the patent of another.

Second, if granted, the reissue patent extends only for the life of the original patent. It in no way extends the term of the original patent, but endures only for the remainder of the unexpired part of the term of the original patent.

Third, no new matter can be added to the reissue patent. That is, the reissue application is based on the disclosure of the original patent and cannot be amended to include disclosures not in the original patent.

Fourth, the claims of a patent can only be broadened through the reissue process if the reissue application is filed within two years from the grant of the original patent. Otherwise, the claims must be of the same scope, or narrower. A claim is considered broadened if it is broader than the original claim in any respect.

And finally, the claims of the reissue application must be for the same invention as that of the original patent. The reissue patent cannot be directed to a different invention.

Reexamination

Once a patent has issued, it may be dealt with by many parties. The owner, for example, may desire to use it to enjoin infringement by others, or to obtain income from the assignment or licensing of the patent to others. Potential infringers or licensees, or established licensees, may wish to consider the strength of the patent and determine whether or not it is in fact valid, all without resorting to expensive litigation.

In both camps of interest, one factor which can affect the strength of a patent is the existence of *prior art* not considered by the U.S. Patent and

Trademark Office when the patent was first examined. Prior art, such as patents or other publications, may show that the patent in question was not novel, or was obvious, and thus should not have been granted.

Where prior patents or printed publications not previously considered by the U.S. Patent and Trademark Office exist, any party may cite them to the U.S. Patent and Trademark Office, describe their pertinency, apply them to the claims of the patent in question, and request that the patent in question be reexamined.

Within three months following the request for reexamination, the U.S. Patent and Trademark Office will determine whether the request raises a "substantial new question of patentability" 37 CFR 1.525(a) affecting any claim of the patent. If such an issue is raised, the patent in question is reexamined to resolve the question.

During this proceeding the patentee may amend his claims and present information in support of patentability. Once a decision is reached, a certificate is issued, either confirming patentability of any claim determined to be patentable, cancelling any claim determined to be unpatentable, and incorporating any proposed or new claim determined to be patentable.

Several aspects of reexamination are notable. First, as opposed to a reissue proceeding, reexamination may be requested by either the patent owner or by any other party.

Second, while the current U.S. Patent and Trademark Office fee for reexamination is significantly greater than that for filing a reissue application, the legal services involved in conducting a reexamination may be significantly less than those required in a reissue proceeding, or in a litigation where the same issues of validity are contested.

Third (and like reissue patents), the term of any reexamined patent is the same as that of the original patent. The term is not extended beyond the term originally granted.

Fourth, while any person can request reexamination, the participation in the proceeding of any person who is not the patentee is limited to filing the request, and a reply to any response the patentee might make to the request prior to the reexamination determination by the U.S. Patent and Trademark Office.

Accordingly, a patentee may wish to test the water respecting the validity of his patent against newly discovered prior art before entering a costly litigation, or to strengthen the patent in a licensing negotiation against prior art presented by an alleged infringer or potential licensee, he may request reexamination. On the other hand, a party concerned about infringing a patent, or faced with paying a license fee, may wish to request reexamination to bring the question of validity over previously nonconsidered prior art to the attention of the U.S. Patent and Trademark Office, thus avoiding the expense of defending a litigation or paying a royalty.

There are many other advantages, disadvantages, and detailed and complicated considerations respecting the use of the reissue and reexamination proceedings. The advice of skilled patent counsel is a must.

9

Chemical Inventions and Patents

Many patent practitioners contend that the preparation and prosecution of chemical patent applications is an area of specialty unto itself. For those readers who likely will never have occasion to deal with chemical inventions, there is merit to reading this chapter because some of the topics covered have applicability to other areas as well. An effort has been made in covering this area not to duplicate materials covered elsewhere.

What is possibly patentable in the area of chemical inventions extends to new compounds, mixtures of chemical compounds, processes of making chemicals, processes of treating materials, new products utilizing chemical compounds, e.g., a new type of plastic-coated paper, and new uses of old chemical compounds and mixtures. The latter class is perhaps not as easily understood and stems from one section of the patent laws which permits the patenting of a new and unobvious use of old chemicals or chemical mixtures. For example, if one patents a mixture of old chemical compounds and discloses as the use thereof the removal of rust from ferrous materials, the patenting of the use of the same mixture of chemical compounds as a stabilizer in polyethylene plastics would be possible if such were new and unobvious to one of ordinary skill in the art. The form of patenting would be by process claims.

Before patenting a chemical invention, the patent lawyer and the engineer should carefully discuss the pros and cons of patenting the invention and thus publishing it versus attempting to protect it as a trade secret. For example, some chemical processes or formulas can be used in secret with competitors not being able to ascertain their identity. On the other hand, once a competitor learns of the identity of the trade secret through *fair means*, he is entitled to practice it also.

Assuming that patent protection is desired, the starting point is again disclosure of the invention in detail to the patent lawyer. Depending on the amount of experimental work that has gone before, this disclosure will probably result in a detailed discussion of the invention to establish:

1. Chemical equivalents that would possibly also work
2. The effect of varying the reaction conditions
3. The possible elimination or rearrangement of various process steps
4. The magnitude of any improvements obtained
5. The utility of the invention
6. Any relevant prior art

One objective of this initial discussion between patent lawyer and engineer is to explore and map out the parameters of the invention. Any added work that must be done in order to define those parameters as well as the additional work necessary to provide a full disclosure should be discussed. This is extremely important because with chemical inventions the breadth or scope of protection that will be obtained is almost certainly going to be commensurate in scope to the patent disclosure.

Having decided what the invention is, a patentability search is certainly the usual course to follow for the reasons heretofore given. The patent lawyer's typical search of U.S. patents can be complemented by a research person's search through *Chemical Abstracts*. Literature references or foreign patents which are unavailable through a routine search at the U.S. Patent Office can often be located through *Chemical Abstracts*

Also of assistance in making the search are several computer search services covering all issued U.S. patents for the last 20-25 years. Moreover, many nonpatent, technical data bases are now easily accessible. Some data bases include only abstracts while others include the complete text of the technical publication. Such can provide a valuable back-up tool in many searches. In many instances the search will suggest other areas for further exploration, e.g., by suggesting additional chemical equivalents that should be tried or other possible end uses for the product.

As mentioned earlier, the scope of the patent will probably be dependent upon the scope of the disclosure. In other words, the more examples defining the parameters, the broader the patent. "Paper examples," i.e., ones that are conceived but never tried, are permissible. However, if they turn out later on, upon duplication, to be inoperable, they can damage or destroy the patent.

The patent laws require that an inventor, at the time of filing, set forth in the application what he then considers to be the best mode or preferred embodiment for practicing the invention. This requirement is more strictly applied in chemical cases than mechanical cases and is frequently urged as a defense in a patent infringement action involving a chemical patent. Care should be exercised in determining the *exact* details of the preferred embodiment. These must be disclosed even though certain details appear to be minor or readily apparent to those in the particular art. It is better practice to disclose them than to have to explain at a later date why they were omitted.

As to what must be given as a disclosure in order to obtain the broadest possible patent coverage, generalizations are difficult but certain guidelines in context with hypothetical inventions can be generalized. If the invention pertains to making a new compound X and the patent lawyer is told that the engineer reacts A and B in the presence of sodium hydroxide at a temperature of 150° F with the constant removal of water, the engineer should review the experimental work to ascertain such parameters as

1. What substitutes there may be for A and B.
2. What proportions of A and B can be used and what effects are produced by varying the proportions.
3. The reason the reaction has to be carried out under alkaline conditions; whether acidic conditions may be used; whether only alkaline conditions can be used; and what other basic materials can be used and in what amounts.
4. What other temperatures, if any, can be used.

5. What happens if water is not removed during the reaction.
6. How certain is it that compound X has been accurately identified.

These foregoing types of questions are usually asked before the search is made but definitely should be explored before the application is filed. Understandably, many inventors, after making an invention of the type under consideration, do not explore these questions. After all, they have achieved the result sought and have found reproducible conditions. Alternative ways may be of very little concern from a business standpoint. From a patent standpoint they are important because filing on only the one example could result in limited patent coverage. Of course, the doctrine of equivalents may be applied in any litigation to extend the coverage, but it is far better not to have to rely upon this doctrine and instead obtain broader coverage at the outset.

Continuing with the hypothetical, let's assume that the discussion shows that A^1, A^2, and A^3, members of a larger class of compounds, also have been tried and work. Examples of these three will be included with the objective being to obtain coverage of the class as a whole. If the class is large, other members will probably have to be tried and given as actual examples in order to convince the Patent Examiner that coverage of the entire class is justified.

If the reaction mixture of A and B has been tried at 50% by weight of A and 50% by weight of B, the proportions of each should be varied to determine what the technically operable ranges are. Examples at each end of the range and in the middle should be submitted as well as the preferred range. In many instances such information will be presented in the form of a table, as for example:

Compound	Range in % by Weight	Preferred Range in % by Weight
A	20-80	40-60
B	20-80	40-60

So, too, will temperature ranges be explored, the ranges established, and examples provided.

In the case of many reactions, optional ingredients can be added to satisfy certain objectives. Such ingredients, although not absolutely required to practice the process, will usually be disclosed, especially if their inclusion constitutes the preferred practice of the process. The same is true of a mixture of chemical compounds, e.g., a dishwashing detergent. Dyes or antitarnishing agents may not be required to achieve operability but their additive effect may be important.

The criticality of ingredients, the possible inclusion of any compound, or the exclusion of compounds is usually apparent from the claims of the patent or application which will include either the words "comprising" or "consisting" or the phrase "consisting essentially of." The first means the claimed invention may include any additional compound, the second means it can include no other materials, and the third means the claimed invention is open to the inclusion of materials which will not substantially affect the claimed invention.

In regard to new chemical compounds, the compound will usually be claimed in terms of its chemical formula or structure. Care should be exercised in determining it. Since some chemical compounds cannot be analyzed with any degree of accuracy, it is permissible to claim such in what are referred to as product-by-process claims. This means what it implies. For example, "A new composition of matter produced by the reaction of A and B, under alkaline conditions, at a temperature of 150°F, with the constant removal of water."

Frequently, in defining a chemical invention, use is made of various types of chemical or physical tests, e.g., X-ray diffraction. Care should be taken to insure that the tests are scientifically valid and reproducible. This is especially true where they will be used to establish a future case of infringement.

There is no requirement that the application set forth the theory thought to explain the invention. If it is done, it is usually couched in such phrases as "it is believed," "while not fully understood," and so on. If such is inserted in the application, it should form no part of the claims since in that case an erroneous theory may render the claims invalid. In contrast, an erroneous theory in the specification will probably have no effect on the validity of the claims except for its embarrassment to the inventor.

In the detailed explanation of the invention in the specification, reference cannot be made solely to the trade name or trademark of a chemical compound to identify a compound. Such does not provide an adequate disclosure since the exact product sold under the trademark or trade name may be changed by the manufacturer at will. Although sometimes it may be difficult to ascertain the chemical nature of the material utilized, a definition other than the trademark or trade name must be supplied. Sometimes this can be supplied in terms of referring to a patent under which the product is manufactured.

Having drafted the specification to permit the broadest possible scope of protection in view of the prior art, the claims should be carefully drafted to coincide with the disclosure. Many patent lawyers draft the claims before they write the specification. Claims which are too broad and have to be amended during the prosecution can be difficult to enforce because of what is termed "prosecution history estoppel." To give an example, suppose the application claims as filed recite 20-80% of ingredient A, the Patent Examiner rejects that on the basis of a prior art reference showing 30% A, and in view thereof, they are amended to cover 40-60% of A. It would be difficult and probably impossible to convince a court that an accused product having under 40% A would constitute an infringement. It would have been less damaging to present claims to 40-60% initially and obtain their allowance without amendment.

The engineer's assistance and participation should not end with the filing of the application. The patent lawyer should be advised of new developments in the area, both successes and failures, so that in consultation with the engineer the application can be prosecuted accordingly and continuation-in-part applications filed if needed. Likewise, the engineer should participate in the preparation of arguments to be used in any response that is to be filed to the Patent Office rejection. The participation should at least include a review of the prior art cited against the claims to provide the patent lawyer with any distinguishing features of the claimed invention. It may frequently include the preparation of factual affidavits to refute the Examiner's opinions.

During the whole development and patenting procedure, complete, accurate, and witnessed laboratory notebooks *must* be maintained. See

Chapter 4. Each test recorded therein must be witnessed by one other than the person performing the experiment. Patent interferences have been won or lost due to the presence or absence of such records in spite of who may have in fact been the true inventor. Similarly, properly kept notebooks may provide proof of early conception and reduction to practice, which proof may be used to eliminate a prior art reference. The notebooks may also provide answers to questions of inventorship as well as ownership.

10

Biotechnology and Patents*

Although biotechnology is a relatively new term of art, biotechnological inventions are not new to patent attorneys or the United States Patent and Trademark Office (PTO). For more than a century, the PTO has been issuing patents drawn to what we now think of as biotechnological inventions. For instance, in 1873, the PTO issued U.S. Patent No. 141,072 to Louis Pasteur for "pure yeast" which was free of pernicious germs and used in the brewing of beer. Only four years later, the PTO issued U.S. Patent No.. 197,612 to William Cutler for a "vaccine virus" made from crushed pustules mixed with fluid lymph and employed in vaccinating persons against smallpox. Since those early years, the PTO has issued a plethora of patents directed to all types of biotechnological inventions, including *living matter per se*!

---

*This chapter is contributed by Peter J. Manso. Mr. Manso is currently with the patent law firm of Wood, Herron & Evans in Cincinnati, Ohio. He received his B.S. degree in biology/chemistry from the Florida State University in 1976. He also received his B.S. degree in pharmacy from the University of Maryland in 1979, and his J.D. degree in law from the University of Cincinnati in 1982. Mr. Manso is currently a registered pharmacist in the state of Virginia and admitted to practice law before all state courts in Ohio. Mr. Manso is also admitted to practice before the United States Patent and Trademark Office and has been specializing in the practice of biotechnology and pharmaceutical patent law since receiving his law degree in 1982.

In spite of the PTO's early involvement in the biosciences, patent protection has not always been available for living matter per se. In fact, up until 1980, the PTO considered living matter per se to be unpatentable. However, in 1980, the United States Supreme Court, in the landmark decision of *Diamond* v. *Chakrabarty*, 447 U.S. 303, 100 S.Ct. 2204, 206 U.S.P.Q. 193 (1980), created a new era with respect to the patenting of biotechnological inventions. There, the Supreme Court ruled that living matter per se constitutes patentable subject matter under the general patent statute, 35 U.S.C. §101. The significance of this decision is dramatically underscored by virtue of the fact that, for the first time, after nearly two hundred years, the Supreme Court interpreted the general patent statute to cover living matter per se.

In *Diamond* v. *Chakrabarty*, the inventor filed an application for patent drawn to a genetically engineered bacterium from the genus *Pseudomonas* which was capable of breaking down crude oil. The novel bacterium contained at least two stable energy-generating plasmids that provided separate hydrocarbon degradative pathways. Because the inventor claimed the live, genetically engineered bacterium per se rather than in combination with culture medium, the Supreme Court was confronted with having to decide whether a live, man-made microorganism qualifies as patentable subject matter under 35 U.S.C. §101. In considering this question, the Supreme Court noted in the legislative history of 35 U.S.C. §101 that Congress clearly intended for it to be construed comprehensively to include as patentable subject matter "anything under the sun that is made by man." Judging the invention in this light, the Supreme Court held that the live, man-made bacterium engineered by Chakrabarty clearly constitutes patentable subject matter under 35 U.S.C. §101.

At about the same time, the Court of Customs and Patent Appeals in *In re Bergy*, 596 F.2d 952, 201 U.S.P.Q. 352 (C.C.P.A. 1979), held that an invention drawn to a "biologically pure culture" of a certain microorganism *Streptomyces vellosus* constitutes a man-made product which qualifies as patentable statutory subject matter under 35 U.S.C. §101.

Since the *Chakrabarty* and *Bergy* decisions, the PTO has significantly expanded 35 U.S.C. §101 to include man-made plants and higher nonhuman lifeforms as patentable subject matter. In *Ex parte Hibberd*, 227 U.S.P.Q. 443 (Bd. Pat. Appl. & Inter. 1985), the PTO held that, in addition to the Plant Variety Protection Act under 7 U.S.C. §2321 *et seq* and the Plant Patent Act under 35 U.S.C. §161 *et seq*, the general patent statute includes as patentable subject matter biotechnological inventions directed to maize plant seeds, maize plants per se and maize plant tissue cultures. In *Ex parte Allen*, 2 U.S.P.Q. 2d 1425 (Bd. Pat. App. & Int. 1987), the PTO held that the general patent statute also includes as patentable subject matter a certain Pacific polyploid oyster. Following the *Allen* decision, the PTO advised the public in a notice dated April 7, 1987, that nonnaturally occurring nonhuman multicellular organisms, including animals, are patentable subject matter under 35 U.S.C. §112.

Statutory Requirements for Patentability

Today, in the aftermath of the epochal *Chakrabarty* and *Bergy* decisions, it is abundantly clear that the PTO considers any invention in the field of biotechnology, whether living or not, to constitute patentable subject matter under the general patent statute. However, like any other invention, a biotechnological invention must conform to the statutory requirements for patentability. The PTO cannot issue a patent on a biotechnological invention unless it is *new, useful* and *nonobvious*. These statutory requirements of patentability are set forth in Title 35, §§101, 102 and 103, of the United States Code and discussed in Chapter Three.

Critical to the novelty requirement for biotechnological inventions under 35 U.S.C. §§101 and 102 is that such inventions cannot exist naturally, i.e., they cannot be "products of nature." Rather, biotechnological inventions must owe their unique existence solely to human ingenuity. This man-made requirement is essential to the patentability of biotechnological inventions and is based upon the idea that since naturally occurring products are not new, they are free to all people and reserved exclusively to none. The famous Supreme Court case of *Funk Bros. Seed Co.* v. *Kalo Inoculant Co*., 333 U.S. 127, 68 S.Ct. 440, 76 U.S.P.O. 280 (1948) illustrates this point. In that case, the inventor had discovered that there were several species of bacteria of the genus *Rhizobium* which existed together in nature without exerting a mutually inhibiting effect on one another. The inventor used that discovery to make a single inoculant containing these different species of bacteria for inoculating several different leguminous plants for which the PTO issued a patent. In holding the patent invalid, the Supreme Court found that the inventor's inoculant did not alter the natural existence of the bacteria in any manner whatsoever. The Supreme Court observed that the inventor's inoculant did not produce any new bacteria, it did not change any of the existing bacteria, it did not enlarge the range of their utility, and it did not change the form in which they naturally existed. The Supreme Court therefore concluded that the inoculant was nothing more than a "product of nature" and hence unpatentable.

Thus, if a biotechnological invention is drawn to a naturally occurring microorganism unisolated from its natural environment, it will not qualify as patentable subject matter since the microorganism in this form is not novel. Such a naturally occurring, unisolated microorganism owes its existence to the handiwork of nature rather than man. On the other hand, if a biotechnological invention is directed to a nonnaturally occurring microorganism, such as one which has been genetically engineered, or a naturally occurring microorganism which has been isolated from its natural environment, it will fall within the novelty requirement since this type of microorganism owes its existence to the sole creation of man.

The *useful* requirement under 35 U.S.C. §101 typically is not a concern to the patentability of biotechnological inventions as long as their utilities are self-evident or sufficiently described in the applications for patent. In those instances, however, where the biotechnological inventions are asserted to have "incredible utilities,"such as a product which is a cure for all cancers, the PTO may be justified in requiring substantiating evidence to be sure that such inventions comply with 35 U.S.C. §101. Regardless of what utility is alleged, commercial success is not required for patent purposes under 35 U.S.C. §101.

In addition to the *novelty* and *utility* requirements, there is the requirement that biotechnological inventions must be *nonobvious* under 35 U.S.C. §103. By *nonobvious*, it is meant that the claimed biotechnological invention for which a patent is sought cannot be obvious to one of ordinary skill in the particular field of biotechnological art to which the invention pertains. This requirement of *nonobviousness* is initially determined by the PTO in its examination of an application for patent.

In determining whether a biotechnological invention is obvious, however, the PTO cannot rely upon an "obvious to try" standard to find obviousness: the two are not the same. The mere fact that it is "obvious to try" something in view of what has been taught in the field does not mean that the invention resulting therefrom is obvious. This is especially true where the prior art contains no suggestion as to how the invention might be accomplished or basis for reasonable expectation that beneficial results will be achieved by proceeding along the lines taken by an inventor. To illustrate this point, the Court of Appeals for the Federal Circuit in *Hybritech., Inc.* v. *Monoclonal Antibodies, Inc.*, 802 F.2d 1367, 231 U.S.P.Q. 81 (Fed. Cir. 1986), held that the district court improperly relied upon the "obvious to try" standard to render obvious an invention drawn to a sandwich assay involving the use of monoclonal antibodies. The court noted that, although the prior art taught polyclonal antibody sandwich assays and monoclonal antibodies, the prior art provided only an invitation to try monoclonal antibodies in sandwich assays and failed to suggest how such sandwich assays might be accomplished. The court therefore concluded that the monoclonal antibody sandwich assay was unobvious, even though it might have been "obvious to try" monoclonal antibodies in sandwich assays.

Thus, engineers and scientists are cautioned to avoid making premature decisions regarding the obviousness of their biotechnological inventions without first consulting with a patent attorney who is knowledgeable in this field.

With this background in mind, what is patentable today in the biotechnological field extends to *any invention created under the sun which is made by man and conforms to the statutory requirements for patentability*. Biotechnological *product* inventions therefore can include such things as proteins, antibodies, intracellular components of plant and animal cells, such as DNA fragments, DNA constructs, DNA promoters, plasmids, vectors, RNAs, ribosomes, chloroplasts, mitochondria and golgi bodies, and living matter per se, such as cell lines, fused cells, plant seeds, tissue cultures, microorganisms, plants and nonhuman animals. Biotechnological *process* inventions can include such things as processes for sequencing DNA, RNA or proteins, processes for genetically manipulating cells, plants or animals, processes for recovering proteins produced by cell lines or animals, processes for detecting and characterizing mutagenic agents, processes for culturing tissue or cells, and processes for diagnosing or detecting biological states.

In addition to these statutory requirements for patentability, an application drawn to a biotechnological invention for patent must satisfy the requirements of 35 U.S.C. §112. Under this statutory section, the application must contain

1. A written description of the invention,
2. A full and clear teaching which enables others to make use of the invention, *and*
3. A teaching which sets forth the best mode for accomplishing the invention (see Chapter six).

In drafting an application for patent in the field of biotechnology,engineers and scientists should work very closely with patent attorneys to ensure that the requirements of 35 U.S.C. §112 have been met. This is especially important in view of the fact that the *written description* requirement is separate and distinct from the *enablement* and *best mode* requirements of 35 U.S.C. §112. The mere fact that an invention is described in an application for patent does not mean that the enablement and best mode requirements have been met. For example, if a biotechnological invention is drawn to a novel, live man-made bacterium, it will be necessary to not only characterize the bacterium, but also to provide a teaching as to how the bacterium can be *made* and *used* by others. Moreover, if the invention includes a preferred bacterium, it is necessary to disclose the preferred bacterium. If the invention is also concerned with novel methods of *making* and *using* the bacterium, it is likewise necessary to describe the best mode contemplated by the inventor for making and using the preferred bacterium.

It is not necessary, however, under the best mode requirement to identify which bacterium, if more than one, is the most preferred bacterium or which methods are the most preferred methods. Rather, all that is necessary to comply with the best mode requirement of 35 U.S.C. §112 is to *set forth* in the patent application the most preferred embodiment(s) known to the inventor(s) at the time the application for patent is filed. If, however, the preferred embodiment(s) *cannot* be readily determined from the patent application as originally filed, this form of disclosure may violate the best mode requirement of 35 U.S.C. §112. See *Ernsthauser* v. *Nakayama*, 1 U.S.P.Q. 2d 1539, 1549 (Bd. Pat. App. & Int. 1986).

In those instances where a certain microorganism or cell line is essential to the practice of a biotechnological invention, it is necessary to precisely characterize the microorganism or cell line in the application for patent in accordance with 35 U.S.C. §112. In the case where the particular microorganism or cell line is well known and readily available to the public, and there is no reasonable basis to believe that such microorganism or cell line will cease to be available during the term of the patent, a complete identifying description in the application is sufficient. However, when the biotechnological invention is concerned with a novel microorganism or cell line which is unknown and unavailable to the public, it is necessary to take additional steps to satisfy the requirements of 35 U.S.C. §112.

An acceptable procedure for meeting the requirements of 35 U.S.C. §112 with respect to an unknown microorganism or cell line is to deposit it with a recognized public repository which affords permanence and availability of the deposit. The purpose of the deposit is to ensure compliance with the requirements under 35 U.S.C. §112 by accomplishing what words are unable to do. That is, a deposit removes any uncertainty regarding the precise characterization of an unknown microorganism or cell line thereby ensuring that others will be able to practice the invention completely. Exemplary of two public repositories acceptable for these purposes are the Northern Regional Research Center (NRRL), 1815 North University Street, Peoria, Illinois 61604 and the

American Type Culture Collection (ATCC), 12301 Parklawn Drive, Rockville, Maryland 20852, though other recognized public repositories are available worldwide.

When depositing a microorganism or cell line to meet 35 U.S.C. §112 requirements, it is essential that

1. The deposit be identified in the application for patent by its deposit number and taxonomic description to the extent available,
2. The deposit be available to the PTO during the entire pendency of the patent application,
3. The deposit be permanently maintained by a recognized repository throughout the term of the issued patent,
4. The name and address of the recognized repository be included in the patent application, *and*
5. There be no restrictions placed upon the recognized repository as to the availability of the deposit upon issuance of the patent.

It is not necessary under U.S. patent practice, however, to deposit the microorganism or cell line in a recognized repository prior to the filing of an application for patent as long as the biological material exists before the application for patent is filed and the deposit is effected and the depository data is inserted into the patent application prior to the issuance of the patent. The case of *In re Lundak*, 773 F.2d 1216, 227 U.S.P.Q. 90 (Fed. Cir. 1985) provides precedent for this principle. In that case, the inventor Lundak filed an application for patent drawn to a novel cell line and the hybridomas resulting from its infusion with lymphoid cells. The unique hybridomas secrete immunoglobulins, which are useful for diagnostic and therapeutic purposes. Before the application or patent was filed, Lundak had deposited the cell line privately in his own laboratory and with his colleagues at the University of California. Lundak also deposited the cell line with a recognized repository, i.e., the ATCC, but not until seven days *after* the filing date of his patent application. Following the formal deposition with the ATCC, Lundak amended his pending patent application to include the ATCC depository data.

The PTO rejected Lundak's patent application under 35 U.S.C. §112 as non-enabling since Lundak failed to deposit the cell line with a "recognized repository" *on or before* the filing date of his patent application. The PTO further rejected Lundak's patent application on the grounds that Lundak had added new matter to the patent application subsequent to its original filing date by virtue of the inclusion of the ATCC depository data.

The Court of Appeals for the Federal Circuit reversed the PTO's rejections. The court held that it is not necessary under 35 U.S.C. § 112 to deposit biological material with a "recognized repository" *prior* to the filing date of an application for patent. The court also held that the insertion of formal depository data into the patent application subsequent to its initial filing date is not new matter. In so holding, the court ruled that it is not critical as to whether the biological material resides in the private possession of an inventor or in the possession of a "recognized repository" as of the filing date of the patent application as long as the biological material is available at all times to the PTO during the pendency of the patent application and to the public following the patent grant. The court therefore concluded that, since the Lundak cell line was available at all times to the PTO and the public by virtue of the fact that it was deposited in a private laboratory at the University of California *prior* to the filing date of the patent application and with the ATCC before the application for patent would issue, respectively, and since the pending patent application was amended to include the ATCC depository data, the requirements of 35 U.S.C. § 112 had been met.

Thus, following the *Lundak* decision, it is permissible under U.S. patent practice to delay the formal depositing of biological material with a recognized repository until after the filing date of an application for U.S. patent without forfeiting the original date upon which the U.S. patent application is filed. The formal deposit, however, must occur *prior* to the issuance of a patent application into a patent and such formal depository data must be inserted into the pending application. In addition, the applicant for patent must file with the PTO a verified statement declaring that the formally deposited biological material is identical to that originally disclosed in the patent application and that it existed before the application for patent was filed.

Unlike U.S. patent practice, however, foreign patent practice requires that the biological material must be deposited with a recognized repository *prior* to the filing date of an application for patent in order to secure as the filing date, the initial date upon which the patent application is filed. Thus, if U.S. and foreign patent protection will be sought, the biological material should be deposited with a recognized repository before any application for foreign or U.S. patent is filed to ensure that subsequently filed corresponding patent applications will receive as their filing dates, the initial date upon which the first foreign or U.S. patent application was filed (see Chapter sixteen for further discussion on foreign patents).

The deposit requirement under provisions of 35 U.S.C. § 112 is not limited by any means to only unknown and unavailable microorganisms or cell lines.

It is intended to apply to any biotechnological materials that are not readily reproducible from their written descriptions, such as novel hybridomas, plasmids, cloning vectors and the like. In some instances, the deposit requirement may apply to higher-life forms, such as plants and nonhuman animals. At this time, however, the PTO has not established an acceptable procedure for depositing such higher-life forms. Nonetheless, the PTO has informally advised that a deposit of the lowest common denominator of a higher-life form, e.g., the sperm egg, fertilized egg or embryo, will satisfy the deposit requirement of 35 U.S.C. §112. The ATCC accepts embryo deposits, preferably in the blastosphere stage. The ATCC also accepts plant tissue culture deposits and plant seed deposits.

Patent Term Extension

On September 24, 1984, Congress enacted the Patent Term Restoration Act which permits the term of a patent to be extended beyond its normal 17 year term. The objective of this Act is to partially restore to the life of a patent that period of time which was lost while the *claimed product* was undergoing review by a federal agency. The statutory requirements for patent term extension can be found in §156 of Title 35 of the United States Code.

Under 35 U.S.C. §156, a patent is eligible for extension if it *claims a product* subject to regulation under the Federal Food, Drug and Cosmetic Act. The term *product* as defined in 35 U.S.C. §156 refers to human drug products, medical devices, food additives, and color additives. A patent which claims a *method* of using or manufacturing a *product* also qualifies for extension. Examples of patents in the biotechnology field which might qualify for term extensions are those which claim *products* or *methods* of manufacturing or using *products* produced by recombinant DNA technology, cell lines, fused cells, transgenic animals, etc.

In addition to the *product* requirement, the patent extension statute is replete with other complex requirements that are beyond the scope of this chapter. Nonetheless, if a patent does qualify for term extension, the time of extension cannot exceed more than 2 or 5 years depending upon the particular patent in question. In either case, however, the total term of an extended patent following completion of the regulatory review by a federal agency cannot exceed 14 years. In the case where a patent has at least 14 years still existing on its original term following the regulatory review, the term of that patent can never be extended.

Patentability of Nonhuman Animals

Recently, the highly controversial issue regarding the patentability of transgenic animals has surfaced. This has come about in the aftermath of the *Allen* decision and the PTO's public notice wherein the PTO concluded and confirmed, respectively, that nonnaturally occurring higher life-forms, such as transgenic nonhuman animals, qualify as patentable subject matter under 35 U.S.C. §101. Apparently, the PTO has adopted the view that the ancillary issues surrounding the patentability of transgenic animals, such as ethical, moral,emotional, environmental, and religious issues, are not the concern of the PTO and should be resolved by either Congress and/or the courts. In response, there appears to be growing public opposition to the PTO's position, and an attempt is currently being made to persuade Congress to impose a moratorium on the patenting of transgenic animals. Also, legislation has been recently introduced into both the House of Representatives and Senate which, if passed, would prohibit the patenting of animals.

At this time, it is too early to predict how this highly controversial issue will be resolved. Nevertheless, since the PTO, in the *Allen* decision, relied heavily upon the pronouncements in the *Chakrabarty* decision nnd the legislative history of 35 U.S.C. §101, and has excluded human life-forms from patentable subject matter obviously in view of constitutional considerations, it may be that the courts will follow the PTO unless Congress elects to amend 35 U.S.C. §101 or create a new statutory section which expressly excludes man-made higher life-forms as patentable subject matter.

In the meantime, however, the PTO took a historical step on April 12, 1988, by issuing the first animal patent ever, U.S. Patent No. 4,736,866 (Appendix F). This landmark patent was granted to scientists of Harvard University for a *transgenic nonhuman mammal* containing a recombinant activated oncogene sequence integrated into the genome of the mammal. Since the novel transgenic mammal and its ancestors are more likely to develop malignant neoplasms, they are considered to be useful in cancer research.

With respect to this controversial issue, it is worth noting that patents represent nothing more than a form of exclusive ownership for a limited period of time, and that people have been owning and selling animals for centuries. Moreover, animal breeding is not new. Geneticists, farmers and breeders have been breeding and cross-breeding animals since virtually the beginning of agriculture to obtain the best possible breeds. In this light, it is difficult to see how the patenting of genetically engineered, nonhuman higher life-forms conflicts with either of these accepted, traditional practices.

Trade Secret, Copyright, and Ownership

Biotechnological inventions also impact upon other areas of law such as trade secret, copyright and ownership law. With respect to trade secret law, it provides an alternative to patents for protecting valuable and sensitive biotechnological information. In some instances, trade secret protection can be even more valuable than patent protection since a patent is granted for a period of only 17 years, whereas a trade secret can last theoretically indefinitely. Thus, the patent attorney and engineer or scientist should carefully consider the pros and cons of patenting a biotechnological invention as opposed to attempting to protect it as a trade secret. For instance, if the biotechnological invention is directed to a *process* involving the production of a protein by a particular cell line, it may be advantageous to elect not to patent the *process* if it can be practiced in secret without competitors being able to ascertain its identity. However, if the *process* will become obsolete within the term of a patent, it might be advantageous to file for patent even though such a *process* can be practiced in secret to avoid the possibility that competitors may learn of the *process* through fair or unfair means.

Some writers have suggested that copyright protection may also be available for certain biotechnological inventions. For instance, a biotechnological invention drawn to a DNA sequence may qualify for copyright protection as a "work of authorship" under 17 U.S.C. §102. In order to qualify, however, the DNA sequence must have originated with the inventor and be fixed in a tangible medium of expression, such as a plasmid, library, vector, cell culture, or transgenic animal.

On occasion, living matter extracted from humans, such as hospitalized patients, will be the subject of certain biotechnological inventions. In those instances, complex issues regarding ownership of these types of inventions can arise. The engineer or scientist and patent attorney should therefore work closely together to carefully develop policies in advance which can help to eliminate these types of problems.

11

Patent Interferences

What Is Interference?

An *interference* is a U.S. Patent and Trademark Office proceeding to determine which of two or more parties was the first to invent (and thus which is entitled to a patent), in those cases where the two parties have each filed patent applications claiming substantially the same patentable invention (and thus "interfere" with each other). In addition to interferences between patent applications, the U.S. Patent and Trademark Office may declare an interference between a patent application and an unexpired patent when each contains claims to substantially the same invention, and when the claim in the application was made prior to one year from the date on which the patent was granted.

Interferences are not declared between applications or between an application and a patent owned by the same party except for good cause. Instead, the U.S. Patent Office will require the party owning the interfering patent and/or applications to make its own determination of priority.

After an interference is declared, the question of priority (who made the invention first) is determined by a Board of Patent Appeals and Interferences within the U.S. Patent and Trademark Office. Once the decision of priority is made, the Commissioner of Patents and Trademarks may grant a patent to the party who is adjuged to be the prior inventor.

What You Must Prove to Win an Interference

If anything can be generalized about the complex field of interferences, it is that in order to prevail you must get there "firstest with the mostest" in terms of conceiving and reducing the invention to practice. Each party will try to prove that he was first to "conceive" the invention, i.e., conceptualize it mentally, and that he was the first to "reduce the invention to practice," i.e., build it. A party may also be required to prove in a particular case that he was "diligent" after his conception in reducing the invention to practice, i.e., did not unnecessarily delay building the invention after mentally conceptualizing it.

Reduction to practice of an invention can be established by either "actual" reduction to practice of "constructive" reduction to practice. Your invention is *actually* reduced to practice when it is constructed and operated if it is an apparatus; when it is made if it is an article or composition; or when it is performed if it is a process. Your invention is *constructively* reduced to practice when a patent application covering it is filed in the U.S. Patent Office.

In some cases, to prevail in an interference, it is not necessary to be the first to reduce the invention to practice. The party who is first to conceive an invention, and who thereafter diligently reduces it to practice, will prevail over another party who conceived the invention second in time but who first reduced it to practice. Conversely, if one party is first to conceive the invention but unexcusably delays reducing it to practice, he will lose to a party who conceived the invention second in time but diligently acted to reduce it to practice prior to reduction to practice by the first party. Thus, the first to conceive is not always the prevailing party; you must be diligent in reducing your invention to practice from a time prior to conception of the invention by a second party.

The U.S. Patent Office will assume that the inventions were made in the order of the effective filing dates of the respective patent applications of each party. Accordingly, the burden of proof in an interference rests on the party seeking to establish a different state of facts.

In meeting your burden of proof or in trying to establish your own dates of conception and reduction to practice keep in mind that as far as the Board of Patent Appeals and Interferences is concerned, the testimony of the inventor alone is insufficient to prove his acts; corroboration of his endeavors by others will be required. It is thus extremely important to maintain witnessed records of your conception and reduction to practice dates and to have others involved with tangible proof of such involvement so that you can adequately support your dates and your activities by the records and by the testimony of others. See Chapter 4, "Documentation of Invention to Establish Your Rights Against All Others."

Procedure in the U.S. Patent Office

An interference may be set up as a consequence of becoming aware of a recently issued patent which claims substantially the same invention as that in your pending patent application. You may voluntarily initiate an interference by copying the patent's claims in your application and identifying the patent. Alternatively, the Patent Examiner may become aware of conflicting patent applications and notify one of the applicants, requesting that he add one or more of the other party's application claims to his application (so that one or more common claims can be used as a basis for the interference). The Patent Office may also require you to file affidavits or other proof of your date of invention in order to show that your work predated the effective filing date of the other patentee or applicant. Obviously, if your invention date is after the filing date of the other person's patent application, you are clearly second, and an interference to determine who is first is unnecessary. In such case the other person's patent application will be permitted to issue and your claims will be rejected over it on the basis that it is prior art.

Once there are common patentable claims in each application, or in an application and patent, an interference is formally declared by notice to you and the other party. At this point, each party may inspect the application of the other party, except for materials which contain allegations of dates of invention.

Each party then submits a preliminary statement, specifying the earliest dates each can establish relative to the conception and reduction to practice of the invention. These are mailed to the Patent Office in sealed envelopes by a certain date. Thereafter, the preliminary statement is served on the adverse party. Each party thus files his statement without knowing the dates the other party will assert. This statement forms a basis for the interference. If your critical dates (conception, reduction to practice) are earlier than those of the adverse party and you can support them with suitable proof, then you should prevail.

An interference is not "tried" before the Board of Patent Appeals and Interferences, as in a regular trial before a judge and jury, but on deposition. Affidavits can be used, but the opposing party has the right to cross examine the affiant on deposition. That is, each witness's sworn testimony is taken before a reporter who transcribes every word. Each party has the right to cross examine the opposing party's witnesses. Documents and exhibits of other kinds relating to the critical dates, etc., are introduced together with the transcript of testimony and this evidence is then presented in written form to the Board. The patent lawyer for each side, through his brief and oral argument, attempts to persuade the Board that his inventor's date has been proven to be earlier than the opponent's.

If you have sustained your burden of proof or if the adverse party fails to sustain his burden of proof, congratulations. You win, having proved you were the prior inventor, and the Patent Office awards a patent to you. The claims of the other party which were involved in the interference are effectively lost to him.

If you have lost the interference and you were an applicant for a patent, the claims of your application which were involved in the interference will be rejected by the Examiner. If your application contains other claims which were not directly involved in the interference or designated as "corresponding" to the claims in interference, then you can continue to prosecute them; however, the Patent Office will consider whether they are patentable over the claims involved in the interference which you lost. If you lost an interference in which you were a patentee, i.e., owned a patent, the Patent Office, by reason of a quirk in the law, has no authority to cancel your patent but it will issue a patent to the winning adverse party. Any later attempt by you, the losing patentee, to enforce your patent would be subject to the adverse priority determination.

Derivation—Or "He Stole It From Me"

Occasionally, when parties deal with each other, the problem of derivation arises relative to subject matter which is later incorporated into a patent application of each. Somewhere along the line each party decides that he invented the subject matter and interfering patent applications result. One or both parties will claim that the other party "derived" the invention from him. In these instances, reliance on authentic, witnessed, and fully corroborated testimony or documents is essential to prove your own conception and reduction to practice. Access by others to your work should be carefully documented.

While the foregoing sounds rather simple, an interference is one of the most complex legal proceedings involving patents. The pitfalls are many and good inventions can be lost to the other party because of technicalities. What you should know about interferences is that you should prepare for them, even before you know they exist, by producing and maintaining complete witnessed and corroborated records of conception, diligence, and reduction to practice. This documentation can solve many problems years after when you are faced with proving you were "firstest with the mostest." See Chapter 4, "Documentation of Invention to Establish Your Rights Against All Others."

12

Ownership of Patent Rights

This chapter concerns itself with the two basic questions: First, what are the rights enjoyed by an owner of a patent? Second, how is ownership of a patent or invention, as between an employer and an inventor or employee, determined?

As an owner of a patent, one possesses the right to exclude others from making, using, or selling his invention. In addition, the patent owner has the right to

1. Grant licenses to others to make, use, or sell the invention
2. File a suit for patent infringement if someone makes, uses, or sells the invention without permission
3. Assign or sell the patent to another
4. Transfer the patent to a beneficiary under a will

Rights 1 and 2 are more fully discussed in Chapters 13 and 11, respectively, and a discussion of them will not be duplicated here.

The right to assign a patent or an application for a patent is specifically provided for in the patent laws. The assignment should be recorded in the

Patent and Tradmark Office. If it is not, then the assignment may be treated as void against any subsequent purchaser for consideration without notice of the prior sale, providing the subsequent purchaser records his assignment at the U.S. Patent and Trademark Office.

The right to transfer a patent or an application by means of a will stems from the patent laws which provide that patents have the attributes of personal property.

Patents or applications for patent may be jointly owned by two or more persons, corporations, or other legal entities. Under such circumstance, each owner owns an undivided share of the whole and enjoys the right to license the patent or application or to sell his share without permission from the other owner(s). In fact, a joint owner of a patent or an application for a patent is free to deal with it as he wishes and without permission of the other joint owner. For example, absent an agreement among the joint owners to the contrary, a joint owner can license or sell the patent to a third party and retain all the royalties or the sale price as the case may be without accounting to the other joint owners. Obviously, this can raise problems if the joint owners have diverse interests. If the joint owners do not wish this result, they must agree ahead of time that rights to the invention will only be disposed of on a joint (versus individual) basis and that all revenues obtained will be shared among the joint owners.

Agreements to assign future inventions, future patent applications, and future patents are *normally* enforceable. Such agreements are frequently included in employment agreements. From the discussion which follows relating to ownership rights arising out of the employment relationship in the absence of an agreement, it will be seen that potential controversies over ownership interests can best be avoided by prior agreement. In fact, and as will become evident, because of the diversity of legal opinions in this area, it is usually preferable to settle this question before the invention is made, by written agreement, rather than to litigate afterwards.

Employment agreements typically include various types of provisions relating to the circumstances under which the inventions made by employees must be assigned. Such agreements usually include (1) definitions of the areas of technology in which inventions must be assigned to the employer; (2) under what circumstances assignment in the defined area(s) of technology must be made; and (3) the time period, if any, after employment has terminated during which assignment must be made. These general types of provisions will now be discussed.

The areas of technology in which assignment is required are usually related in some manner to the existing business of the corporation. For example, assignment is required in many typical employment agreements where the invention is made "in any business line of the employer," "in the course of the employment," "in the area of bed springs" if the employer is in the bed spring business, "in any area in which the employer is interested," or "in any area of assigned work." Some employment agreements extend to areas outside of the present business of the employer, e.g., to all inventions that may be of interest to the employer; however, there is some likelihood that such a provision may be held unenforceable in court on the basis that it is contrary to public policy.

When an assignment of an invention is required by the terms of an employment agreement, it usually makes no difference that the invention was not made during the normal hours of employment. Such is valid. The reason is obvious because engineering and management personnel are expected to be thinking about their jobs not only during normal working hours while at their place of employment but after hours as well.

The time period during which one may be obligated to assign an invention may extend beyond the term specified in the employment contract if the extended time, in view of all the circumstances, is reasonable. Again, public policy intervenes to invalidate the agreement if the circumstances show the provision to be unreasonable. If the agreement extends only to the employment term, assignment will be required. As sometimes occurs, an employee makes an invention in the area of technology covered by his employment agreement, does not disclose it to his employer, and subsequently terminates his employment and attempts to market the invention on his own behalf. In such cases he can be forced to assign the invention to his previous employer.

A promise to employ a person is sufficient consideration to support a promise to assign future inventions made during the employment. It is sufficient to support assignment of inventions made subsequent to termination of employment term, provided the agreement is reasonable as to area of technology and time. A different situation is presented when an employee who has no written agreement to assign is asked to enter into one. Some courts would require the employer to provide consideration in addition to that of continued employment, e.g., a bonus.

In the absence of an employment contract, written or oral, defining the circumstances under which an employee must assign, there may be an implied duty to assign, *depending upon the particular factual circumstances and the law or judicial decisions which are applicable*. The complexity of this question is evidenced by numerous decisions of the U.S. Supreme Court of rather early vintage and the plethora of the litigation since then. Hardly a month goes by when the question is not litigated! Obviously, very strong questions of public policy are involved which, while they are not always discussed, do enter into the judicial decision-making process.

In order to find a duty to assign, the usual legal controversy centers around the question of what the employee was hired to do. This presupposes that an invention was made by the employee relating to the general business of the employer and made during business hours with company material.

In trying to arrive at an explanation for and answer to this question, two extreme hypotheticals may be posed, the first being the case where company A in the business of making automobiles hires a janitorial employee who, during his cleaning of the assembly line one day, his only assigned duty, views an automobile engine and because of the carburetor problems on his own car conceives how to introduce gasoline into an engine in a much more efficient manner and thereafter reduces the idea to concrete form on his own time using his own materials. The second hypothetical is the case of a private research laboratory which has a large automobile manufacturer as a client, hires as an employee a person with a Ph.D. in automotive engineering with employment experience in automotive fuel systems, and assigns to him as a research project the task of inventing a new automotive fuel system during which he invents the same fuel system as did the janitor. In case 1, few if any courts would require assignment, while in the second case few courts would not. The difference in the two situations which dictate opposite results resides in the job duties of the respective employees and what they were hired to do. The Ph.D. was hired "to invent" and will normally be required to assign; the janitor was not.

The problem with determining the difference between the two situations stems from the definition of "hired to invent." For example, what if one is hired to "improve" an existing product line? Two U.S. Supreme Court cases say this is not "hiring to invent," while loose language of the U.S. Supreme Court in other cases suggests the contrary. And one can find many lower court decisions on each side of the issue.

At first blush it seems harsh to deprive the employer of the fruits of an employee's work when he has been hired to improve. But, in those states where an employee is hired to improve and the employer does not own the invention, the employer is still entitled to make and use the invention without paying the employee and without being guilty of patent infringement should the employee patent the invention on his own as he would be entitled to do. This is because of the judicially created doctrine known as "shop rights." The use of the employer's tools and materials during the time for which the employee is paid gives to the employer this shop right, which amounts to an implied royalty-free nonexclusive nontransferable license, even though the employee owns the invention. Such shop right, however, does not give to the employer the right to license others.

It is better in most situations to eliminate these problems beforehand by entering into an agreement which is fair to both employer and employee.

13

Enforcing Patent Rights

In this chapter we will attempt to develop for you a general outline of patent infringement problems. We will discuss how the infringement conflict arises, how to avoid it, and what happens once the problem arises.

It may be helpful to discuss the infringement problem in terms of the White Company and the Black Company, who are competitors.

It is perhaps important to note that in many situations there are no clear-cut good guys and bad guys but rather two companies operating under the reasonable belief that what each is doing is proper under the circumstances.

How the Infringement Conflict Arises

There are many ways an infringement problem can develop. We will mention a few of them.

Black Company, through its engineers or research and development group, develops a new apparatus for making widgets. They construct the apparatus, find it to work satisfactorily, and introduce it into the market.

Unfortunately, in their ignorance of what White Company is doing, they find they are accused of infringing White's patent on a comparable machine. White Company gives notice of infringement and a lawsuit for infringement is about to begin.

Take another situation. Black Company sees the apparatus that White Company has marketed for manufacturing widgets. Black Company is being ruined by the competition and decides that it needs to duplicate substantially the White Company's machine. They therefore ask their engineers to develop the machine which will do substantially the same job as the White machine and thereafter introduce such a copy or close copy into the marketplace. The White Company naturally takes umbrage at the Black copy of its machine and sends a notice of infringement. Once again, an infringement lawsuit is about to begin.

Let us take still another example. Again, the White Company has begun the manufacture of apparatus for making widgets. White Company takes care to put its patent number on each of the machines which it manufactures. Black Company, seeing the machine, obtains a copy of the patent. Its patent counsel advises for one reason or another that the machine can be substantially duplicated free from infringement. Based on this advice, Black Company begins the manufacture of its infringing machine. Quite understandably, the White Company accuses Black Company of infringing its patent.

We could go on with variants on the same theme but these will suffice to give you a picture. So, before getting into the infringement struggle, let us consider some of the things you might think about to avoid this infringement problem.

How to Avoid the Charge of Infringement

The most important course to pursue in avoiding a charge of infringement is to know ahead of time what patents may apply to your particular development, and here we refer to both expired and unexpired patents. These patents can be obtained through a patent search which is relatively inexpensive—certainly far less expensive than the combined costs of development and a patent infringement suit. Depending upon the complexity of the invention, at today's charges you might be able to obtain an infringement search for anywhere between $1000 and $4000.

A good infringement search will provide much valuable information. The expired patents will show you what can be manufactured free from infringement because, with very limited exceptions, you are entitled to copy anything that is in the expired art.

The unexpired patents will enable you, with the help of a patent lawyer, to evaluate whether or not there are any broad patent claims that will cover the device, apparatus, or the like which you intend to develop and market.

The combined unexpired and expired patents will enable you to get a good feel as to whether the "dominating" (or possibly infringed) patent is a very narrow patent which you can easily avoid by a specific design or whether it is entitled to considerable breadth so as to cover you no matter which way you turn in your development.

Finally, the combination of all of the patents may give you some insight as to where you can alter your prospective design, perhaps moving toward the old expired patents, in order to avoid infringement of the more recent patent.

Thus, having a good knowledge of the existing patent literature on any subject may provide many shortcuts to your design and subsequent marketing of a desirable product free from infringement. See Chapter 5, "Types of Patent Searches and What They Can Do for You."

Before launching an activity that could be accused as being an infringement of a known patent, a company should, by all means, get the *written opinion* of patent counsel that that activity will not infringe the patent. The opinion must be well considered. At a minimum, it must reflect a study of the prosecution history of the patent. The consequences of a failure to have gotten such an opinion could be a decision that infringement was willful and an assessment of treble damages against the infringing company.

Detecting Infringement by Others

This section should not take too long.

Keep your eyes open!

If you have a pretty good understanding of what your company considers to be its proprietary products, as you become exposed to the products of your competitors through reading of the trade publications in your field, possibly studying the *U.S. Patent Office Gazette* where the patents issued every week are digested and classified, or through visiting trade shows, you can observe when the Black Company comes out with a product which encroaches upon your company's territory.

Bringing the encroachment to the attention of your company officials or their patent lawyers will start the ball rolling to determine whether or not your company should be involved in charging the Black Company with infringement and the consequences of it which will appear below.

The Infringement Suit

There are a number of variations on how infringement suits are generated, but we will discuss only the most straightforward type.

In this suite, White Company has determined that there is an infringement of the White Company patent. On advice of counsel, the company officials decide that a suit for patent infringement should be filed.

A complaint for infringement is drafted. This can be a very simple document almost copied out of the forms in the *Federal Rules of Civil Procedure* by which patent infringement suits are governed.

The complaint must be filed in a Federal District Court in the location where the accused infringer, the Black Company, resides or where it has the combination of a regular and established place of business *and* is committing acts of infringement.

After the complaint is filed, the accused must file an answer and any counterclaims. In the normal course, the accused Black Company will not only answer the fairly simple allegations in the complaint but will also counterclaim for a declaration of invalidity of the patent. In this answer and counterclaim, it is not unusual for every conceivable defense to the charge of infringement to be raised. Furthermore, it is not unusual for every conceivable counterclaim to be raised. Here, we refer specifically to charges of fraud on the U.S. Patent Office, antitrust violations, unfair competition, and the like.

With few exceptions, these defenses to the invalidity of the patent and the counterclaims cannot be supported by facts then known to the defendant or by law. Nevertheless, they are raised, perhaps for three reasons.

First, the defendant may hope he will frighten the plaintiff patentee into a modest settlement. Second, if the defendant has pleaded all of these defenses and counterclaims, then they become issues in the case as to which the defendant has a right of discovery. This assumes some importance when he is put to the task of demonstrating that some of the discovery questions which he asks are relevant to the issues in the case. Third, the defendant is either hoping that something will turn up which will justify his defenses or counterclaims or he is afraid that if he has not put them all in his answer and counterclaim he may thereafter be precluded from amending his answer and counterclaim.

Thus, if you are the inventor of the patent in suit, do not be overly concerned about all of these anguished cries of fraud, "ripping off" other inventors, or other acts of malfeasance. When your heart is pure, you will know for a certainty that this is just so much of a preliminary skirmishing in the litigation.

After receiving the answer, the plaintiff patentee files a reply in which he normally denies everything that the accused defendant has written. Upon the filing of these three documents, which are called the *pleadings*, the issues are established and pretrial activities begin.

There may be motions filed to dismiss some of the counterclaims, to strike some of the pleadings that are not justified, or for summary judgment. The motion for summary judgment may be brought when it is clear that there are no factual issues in dispute and the judge can make a ruling on one or all of the issues on the basis of the law.

The major activity in which the litigants are involved prior to the actual trial of the case is discovery. There are four main avenues for discovery.

The first avenue is interrogatories or a set of questions directed to the opponent which must be answered under oath or objected to. The second is a request for the production of documents or things, which must be furnished, or the request objected to. The third is a request for admissions which must be answered or objected to. The fourth involves the taking of depositions. This fourth discovery tactic involves giving notice to the opposition that one of the parties intends to take the depositions of certain named individuals on a certain day. At the appointed hour, the lawyers and the witnesses show up at the appointed place before a notary public who is usually a court reporter, and testimony is taken by direct examination, cross-examination, and redirect examination and is usually recorded. The deposition is similar to testimony at a trial except that there is no judge ruling on the evidence.

The function of discovery, and particularly discovery depositions, is to find out as much as possible about the opponent's case, as well as to develop facts relating to your own case. Sometime during the discovery depositions, the opposition witnesses will drop a few nuggets that are admissions and harmful to the deponent's case. The deposing lawyer will happily seize upon these.

For the most part, however, discovery is to determine as much as one can learn about the opponent's case, so that when the litigants come to trial there will be no surprises.

This discovery process unhappily is time-consuming and expensive. It is not only expensive from the standpoint of the cost and attorney's charges which must be billed to litigants, but also from the standpoint of the litigants' personnel who must spend time searching through documents, answering interrogatories, and getting on the witness stand during the taking of depositions.

The discovery procedure in some cases is unreasonably expensive. There are reasons for this but for the most part none is very good. The only good reason for the very considerable expense of discovery is that there are so many complex issues in the case that many, many hours must be spent pouring over documents, taking depositions and the like which generate large costs. Otherwise, the cause for high discovery costs arise out of:

1. The patent lawyers' being overly protective of their engineers and resisting every effort on the part of the other side to discover facts.
2. Attempts on the part of one side to go far beyond the issues of the case so that the discovery effort is so broad as to require being resisted.
3. Overcautiousness on the part of one or both of the patent lawyers believing that they must leave no stone unturned because the stakes are so high.

Pretrial

Now we are approaching trial, discovery having been completed.

Prior to trial, it is not unusual for the judge handling the case to require the patent lawyers to come in for a pretrial conference. There, the judge attempts to get both sides to admit as many facts as possible and to narrow the issue to be tried so as to shorten the time for trial. The judge may also very well try to cajole the litigants into settling their differences.

A word about settlement. It is always to be favored and explored as early as possible in the litigation. One factor is simply the cost of litigating, and when it is understood that this may be from $50,000 to many hundreds of thousands of dollars, settlement should be seriously considered.

If the opposing side's position is so utterly without merit that under no circumstances is settlement justified, then it may be advisable to fight on in the hope that the judge ultimately will not only decide in your favor but also will award attorneys' fees. The award of attorneys' fees so rarely happens that it is not a major factor to consider in deciding to pursue the litigation to its ultimate conclusion.

In most patent litigations, both sides have some merit in their positions. The outcome could be 50/50. In that situation, it might be highly advisable for the litigants to settle, each litigant realizing that he got less than he had hoped for but, on the other hand, he didn't get completely wiped out.

The Trial

Now let us assume that the pretrial conference resulted in at least some agreed-to statements of facts and it was determined that settlement was not possible.

Insofar as the agreed-to statement of facts is concerned, it probably results in very little more than admissions of the obvious which would take no more than 15 minutes of trial to prove, such as the ownership of the patent, the existence of the accused machine, or the like.

The trial can be a jury trial if it is demanded in time, or it can be a non-jury (bench) trial before a single judge.

Setting aside the subsidiary issues of fraud, antitrust violations, and the like, the infringement trial normally boils down to two simple issues, namely, infringement and validity.

As to the infringement issue, the patentee has the burden of proving there is an accused product, process, or the like which infringes the patent. The defendant, on the other hand, attempts to prove that the patent is not infringed by the accused product, process, or the like, and attempts to prove that the claims of the patent are invalid.

Since the subject matter of the patent is usually rather technical and since the judge trying the case is not normally a skilled engineer, much of the trial will be directed to the introduction of charts, diagrams, models, and the like in order to guide or assist the judge in his understanding of the patent, the prior art, and all of the issues. Both parties will usually use at least one expert witness to help clarify these complex technical matters for the judge.

In all of this, the engineer, assisting the patentee's lawyer, can help a great deal by assisting the patent lawyer to focus on that which is *unobvious* in the patented invention. On the other hand, the engineer who is employed by the accused can help the patent lawyer by demonstrating how *commonplace* or *obvious* all of the features in the patent truly are.

Following the trial of the case, it is not unusual for the judge to request post-trial briefs to be prepared by the patent lawyers on both sides. After receipt of the briefs, it is often many months before the decision is handed down.

After receiving the decision, the losing party has an opportnity to appeal to the United States Court of Appeals for the Federal Circuit. It is not unusual at this time for serious settlement negotiations to be undertaken. If no appeal is taken or if the judge is upheld after the appeal and the patent is held valid and infringed, the losing party may be required to pay damages from the time that it had been given notice of infringement and it may be enjoined from any further infringing activity.

Damages are by law no less than a reasonable royalty. They can be much greater. For example, if winner White Company makes a profit on all of its widgets and if it can show that it would have sold all of the widgets Black Company sold (if Black Company hadn't infringed), White Company could be awarded more than reasonable royalty. It could be awarded all of the profits lost because of the infringing sales. The court is allowed to increase damages, up to three times, if it finds that the infringement was willful.

The defendant can also be awarded damages, trebled, if a court finds that the infringement suit was brought in bad faith and amounted to an anti-trust violation.

Attorneys' fees can be awarded to the prevailing party. Such an award will be made only where the losing party has been guilty of irascible conduct well beyond mere losing of the suit.

The foregoing discussion has contemplated a company infringing in the United States. But what about an alien competitor who manufactures abroad and ships into the United States? That alien competitor can be sued through the process described above. Additionally, an act can be brought before the International Trade Commission (ITC) to stop importation of the accused articles. An advantage of that action is that it is quick. A decision must be made within twelve months. A disadvantage is that it introduces other factors such as the quality of the affected industry in the United States and the injury to that industry that the infringement is causing. But it is an option that should be considered.

14

Getting Around Your Competitor's Patent

In the foregoing chapter we discussed procedure for enforcing your patent against an infringer. Now the shoe is on the other foot. You find yourself in a position wherein your competitor has introduced a patented product that is killing you in the marketplace and you need to produce a comparable product that will not infringe his patent. One of the options is to "design around" the patent. The function of this chapter is to tell you generally how to do it, but this is another area where you really should have the advice of a patent lawyer for the reasons that will become apparent from the discussion below.

Designing around the patent of a competitor is no heinous offense. As a matter of fact, that practice over the years has resulted in many, many valuable improvements and, hence, contributions to the technology of the country. One of the values of our patent system to the country is that it stimulates inventors to create improvements that will avoid infringement. Thus the technology of the country is advanced.

In general, to design around your competitor's patent you must avoid the language of all of the claims of the patent. The claims, as you will recall, are the numbered paragraphs at the end of the written description. Stated another way, if none of the claims of the patent is readable on your proposed product, you *probably* do not infringe the patent.

We emphasize the term "probably" for the very good reason that in the patent law we have a doctrine of equivalents whereby the courts have recognized that minor deviations from the invention as described in the claims will not necessarily avoid infringement.

The courts have also developed a doctrine of prosecution history estoppel, which is somewhat the converse of the doctrine of equivalents. In accordance with the doctrine of prosecution history estoppel, if to obtain the patent, the patentee was forced to amend the language of his claims in order to avoid prior art and thus to obtain his patent, he may be estopped to place a broader interpretation of his claim than that set forth by the amended language.

The courts have had difficulty with the application of these two doctrines, and therefore we suggest the obtaining of the advice of a professional before making a final determination with respect to your product.

By way of example, one of the claims of Patent No. 3,225,299 (Appendix D) is as follows:

> In an apparatus for measuring the speed of a rotating member, a device comprising a housing, a shaft, means carried by said housing for rotatably journaling said shaft, a cyclindrical rotor carried by said shaft, said cylindrical rotor comprising a cyclindrical glass tube and an electrically conductive coating formed on the inner surface of said tube, a first electrode including a plurality of pointed electrode elements directed toward said rotor and being disposed adjacent to said rotor, and a second electrode including a plurality of pointed electrode elements directed toward and spaced from said rotor in an area spaced from said first electrode circumferentially with respect to said rotor, said first electrode being effective to emit electrons and said second electrode being effective to collect electrons previously deposited on said glass tube by said first electrode.

Observe that the claim is limited to "said cylindrical rotor comprising a cylindrical glass tube." If you make the rotor from a material other than glass, will you avoid infringement? That would surely seem to be an easy way to avoid infringement of that claim but beware of infringing those of the remaining claims which are not limited to a glass tube. If the material that you propose to substitute for glass is a recognized equivalent of glass in that type of environment, then the mere substitution of a different material for glass might not be considered to avoid infringement because of the doctrine of equivalents.

A review of the prosecution history, i.e., the proceedings before the U.S. Patent and Trademark Office, may show that in order to get that claim allowed the patentee was forced to be very specific in the recitation of "glass" and argued that "glass" distinguished the invention over the prior art. If you were to find that to be the situation, then the doctrine of prosecution history estoppel would preclude the patentee from asserting that his claim was broad enough to read on a tachometer whose rotor was constructed of a material other than glass.

Another way to get around a patent is to have a validity search made with respect to the patent and to turn up prior art that was not considered by the U.S. Patent Office Examiner and invalidate the patent. This is not always easily done, for we hope that our patent Examiners make thorough prior art searches. However, some patents are mistakenly issued because of the failure of the Examiner to find the most pertinent art, and therefore this possibility must be borne in mind.

Now that you are armed with this background information, you are ready to be exposed to some of the steps which are employed in getting around the patent of the competitor.

1. Obtain a copy of your competitor's patent. The product may have a patent number on it. If not, a search at the U.S. Patent and Trademark Office, particularly the patentee index and assignment records, may produce the competitor's patent. When you have the number of a patent, a copy can be conveniently obtained from a library, or if not, it can be obtained by writing to the Commissioner of Patents and Trademarks in Washington, D.C.
2. Read the patent. The claims of the patent are difficult to understand even to the trained patent lawyer, but without reading the detailed description upon which the claims are based an understanding of the claims is next to impossible. Further, according to the U.S. Patent and Trademark Office rules, the language of the claims should have an

antecedent basis in the detailed description. Therefore, if you are having difficulty in understanding what a particular clause in the claims refers to, refer to the detailed description which should provide the necessary clarification.

Look for one or more features set forth in the claims which you can eliminate from your product or which you can modify in your product. It may be possible to determine at this stage that your proposed product avoids infringement of all of the claims because your proposed product does not respond to any of the claims, and there is no way that the doctrine of equivalents could give the claim sufficient breadth to encompass your product.

3. Obtain the patent references cited by the Examiner. You will find these listed on the front page of the more recent patents and in a section following the claims of the older patents. A study of these patent references may do two things for you. First, they may show you why the claims are limited in a certain way, i.e., you may be able to see that the claims were probably limited in order to avoid one or more of the cited prior patents. Second, the references cited by the Examiner may give you ideas for designing around your competitor's patent.
4. Study the file history. Your study of the amendments that have been made to the claims by the patentee along with the arguments presented to the U.S. Patent Office, all taken in conjunction with the patent references referred to above (which probably generated the need for the amendment and arguments), will demonstrate more specifically why the claims are limited in a certain way. At this stage, you may be reasonably reassured that your proposed product avoids infringement of your competitor's patent.
5. It may be advisable to make a thorough search of the U.S. Patent Office records for prior art which may have been overlooked by the Examiner. See Chapter 5.

As we stated above, you may find prior art that invalidates your competitor's patent, thereby substantially reducing your concern about the possibility of a suit for patent infringement. Perhaps equally important, however, you may find a broad patent that dominates not only your competitor's product but also the product you propose to design. The broad patent may be much more difficult to design around and obviously an awareness of it at an early stage is advisable.

Another value to be gained from the patent search is an exposure to a lot of different products of the type that you are considering designing. Thus, your creative work in designing around the patent may be stimulated. Bear in mind that what is disclosed in an *expired* patent as a general rule may be used free from infringement, and thus if an expired patent discloses a product which you can manufacture in competition with your competitor, the expired patent in most cases provides a "right to use" and gives you all of the comfort which you need in knowing that there is no way your competitor's patent could be successfully asserted against you.

As stated in Chapter 12, obtaining the opinion of competent patent counsel before introducing the product is of utmost importance. Recent decisions of the Court of Appeals for the Federal Circuit suggest that the absence of such an opinion will result in a decision that, if infringement is found, infringement was willful. A determination of willful infringement can form the basis of trebling of damages and awarding of attorney's fees.

15

Exploitation of Inventions: Licensing

Once an invention is made, its owner may embark upon one or more of a number of alternative courses of conduct with respect to it. For example, the owner may decide to do absolutely nothing. He may put it on a shelf in his laboratory, send it to the archives, or lock it up in a closet. The owner may not have the necessary capital, marketing expertise, and like resources required to successfully introduce the product into the marketplace.

Perhaps he has found that someone else has discovered a better way to do the same thing and he believes his own invention cannot compete effectively. Quite possibly the owner himself has subsequently discovered a new and better way to do the same job and wishes to promote the subsequent invention rather than his first one. In large corporate research departments, different inventive solutions to the same problem are often proposed and only the best one exploited.

Sometimes the situation arises that the owner of an invention already has a successful product on the market and in his own mind cannot justify the expense of tooling up and getting a new product into production when the current one seems to be selling just as well as he would expect the new one to sell. In other words, his new invention would do little more than make his existing product obsolete, and as long as he is selling the existing product in sufficient quantity, there is no justification for going to the expense of introducing the new one.

Whatever the reason, sometimes an invention is made and not commercially exploited. Independent of whether an invention is exploited or not, the question of patenting arises. An owner may patent an invention even though he has no intention, present and/or prospective, of ever commercializing it. There are no regulations requiring a patent owner to commercialize a patented product. The invention owner electing not to exploit an invention may also decide not to patent it. Of course, if the inventor abandons, suppresses, or conceals his invention in lieu of obtaining a patent and/or commercializing it, he may forfeit his right to obtain a patent. Moreover, under some circumstances a person who subsequently makes the same invention can obtain a valid patent on it. The second inventor can then assert his patent against the prior inventor who had abandoned, suppressed, or concealed the invention should such prior inventor later decide to commercialize his invention.

Another alternative open to one owning an invention is to start commercializaing it, e.g., by manufacturing and selling products incorporating the invention. If the invention is such that it can be discovered by inspection of the product and the owner wishes to be able to exclude others marketing the same invention, he is well advised to apply for a patent. If he does not apply for a patent and starts selling or offering the product for sale, after one year he has permanently forfeited his right to obtain a patent. Of course, in these circumstances, where the original inventor has not abandoned, suppressed, or concealed the invention, a subsequent inventor cannot obtain a valid patent.

The rights a patent provides to the owner of an invention, namely, the right to exclude others from making, using, or selling the patented invention, are in most cases useful, if not necessary, to maximizing profit from commercialization of products incorporating the invention. However, occasionally an owner may elect to market products incorporating his invention without the benefit of a patent. For example, a company's technology, marketing techniques, and/or channels of distribution may be so formidable that the company can successfully commercially exploit an invention with little or no competition without a patent.

If the invention is of a type that permits its commercial exploitation without discovery of the invention by the consuming public, the owner may elect not to obtain a patent but rather may decide to treat the invention as a trade secret while at the same time exploiting it commercially. For example, the invention may involve a method of heat treating steel for use in making cutlery. If cutlery that has been made according to the process can be sold and the purchaser is unable to ascertain from an examination of the product the process used to make it, as long as the owner can maintain the process a secret and someone else does not independently make the same invention, the owner can market his product without competition from others using the same process even though he has no patent.

As you can see, the owner of an invention may do nothing with an invention and apply or not apply for a patent as he sees fit, or he may go into production and commercialize the invention, doing so with the aid of a patent or on a trade secret basis. There are still other avenues an inventor may elect to pursue.

For example, if the invention is patented he may decide to profit from his invention by transferring some or all of his rights in the invention to others for something of value. He may sell all rights in the invention to another, leaving himself none, for a fixed one-time fee, e.g., $10,000 payable either immediately, at some future date, or in installments over a period of time. Alternatively and/or in combination, payment may be predicated on a royalty basis, e.g., the inventor receiving $150 for each item sold which is covered by his patent. If an inventor, or the owner of the invention if it be someone else, transfers all his rights to an invention, it is known as an *assignment*. Under certain circumstances, payments received in exchange for an assignment receive favorable tax treatment as a capital gain. Thus, an assignment may be preferred if for no other reason than the possibility of obtaining capital gains tax treatment for the proceeds.

If the owner of an invention, patented or not, transfers less than all of his rights, a *license* results. By way of illustration, you will recall that, if a patent has been obtained on an invention, the patent gives the owner thereof the right to exclude others from making, using, or selling the invention. The owner of the patent, therefore, can license another to manufacture his invention, and/or use his invention and/or sell his invention. If the patent owner licenses these rights to only a single person and pursuant to the terms of the license is not permitted to license others, the license is termed an *exclusive license*. However, should the owner license his rights to multiple parties or to one person, reserving the right to license it to others at a future date, the license is termed a *nonexclusive license*. If an exclusive license is granted to another, the patent owner cannot make, use, or sell the invention unless he has expressly reserved the right to do so in the license agreement. As in the case of an assignment, payment for a license may be based on a fixed fee or a royalty or a combination of both.

Thus the owner of an invention may profit from his invention through the efforts of others by granting them an assignment, an exclusive license, or a nonexclusive license in exchange for something of value, usually payment in the form of a fixed fee and/or a royalty. In some cases the item of value received for the license or assignment is not money, but may be, alone or in combination with money, rights in an invention owned by the one receiving the assignment or license. Such a situation in which parties exchange invention rights is called a *cross-license*.

Should the owner elect to treat the invention as a trade secret, he may, in a manner analogous to licensing or assigning a patent, license or assign his trade secret and thereby exploit the invention through one or more additional parties, either as a supplement to his own commercial activities or as a substitute for them. For example, the owner of a trade secret process for heat treating cutlery may transfer the necessary technology to another pursuant to an agreement requiring the recipient to pay certain monies and take the necessary precautions to maintain the technology secret. Depending upon the specific terms of the agreement, the activities of the recipient of the trade secret technology may supplement commercialization by the owner or be a complete substitute for it.

Where a patent has been applied for but the patent has not yet been granted, the owner of the patent application may, in much the same manner as if the patent had been granted, transfer by license or assignment some or all of the rights in the invention to others. When the patent application is granted, the rights granted under the application automatically convert into rights under the patent. Should the U.S. Patent and Trademark Office refuse to grant the patent for some reason—e.g., because it deems the invention to be obvious—the obligation thereafter of the party to whom the rights were transferred will depend upon the terms and provisions of the agreement made at the time of the license or assignment.

Still other approaches to exploiting an invention exist. For example, a person could conceivably make an invention, immediately apply for a patent, and while it is pending and maintained in secrecy by the U.S. Patent Office commercialize the invention as a trade secret, providing it is susceptible of commercialization without discovery. Alternatively, or in conjunction with the above, he could license it to others on the condition that they maintain it as a trade secret until the patent issues, or indefinitely if the patent is not granted. When and if a patent issues, the party receiving the license, termed the *licensee*, will have his license under the patent application transformed into a license under the issued patent.

In some cases, a written instrument evidencing a license or assignment of a trade secret, patent application, or patent is required, such as when a party's obligation under the agreement cannot be completely fulfilled within one year. If it is necessary to have the agreement in writing, it must be signed by the party against whom you wish to enforce the agreement. While there is no legal requirement that a particular license or assignment of a trade secret, patent application, or patent be in writing, to avoid disputes from later arising with respect to the terms the parties have actually agreed upon, it is certainly desirable to have the agreement reduced to writing and signed by each of the parties.

Variations in the terms of a license agreement are virtually limitless, depending upon the needs, desires, and bargaining positions of the parties involved. The license should identify the parties, the date and place where the agreement is made, and the licensed subject matter whether it be a patent application, patent, and/or trade secret. It will also include a definition of the rights being licensed. The license may be nonexclusive or exclusive, with or without the owner (termed the *licensor*) retaining the right to practice the invention. The license may permit the licensee the right only to manufacture, or only to use, or only to sell the invention or any combination of two or more of these rights. The license may be limited to a specific geographic territory, such as certain specified states, countries, market areas, or the like. The license may cover multiple inventions in which case the license is often referred to as a *package license*. The license may cover only certain of the claims of a patent should a patent be involved or only certain specific versions or embodiments of the invention.

The license may include limitations on the quantity that the licensee can produce. For example, a licensee may be permitted to manufacture and sell only 50,000 units of the patented invention per year, or no more than 20% of industry sales. Alternatively, the license may restrict the licensee from utilizing more than one-third of his production capability for manufacture of the licensed item. The license may also contain limitations restricting the licensee to selling only to certain types of customers, such as wholesalers, retailers, jobbers, etc., or be limited to selling to only certain industries. Care must be exercised to ensure that license provisions such as these do not create a patent misuse or violate our antitrust laws (see Chapter 17).

Another principal provision of any license agreement is the monetary consideration paid by the licensee to the licensor. As mentioned earlier, the consideration or value paid by the licensee may be in the form of a fixed sum paid at the time the license is entered into or at some specific date thereafter or it may be payable in periodic payments or installments. Alternatively, the consideration may be variable in nature. For example, the consideration may be in terms of a royalty requiring a payment of 10% of the selling price or 5% of the manufacturing cost. The royalty rate may be fixed for all units made or may vary. For example, the royalty provision may impose on the licensee a duty to pay 10% of the sale price of the first 100,000 units sold each year (or during the life of the license) and 5% on all units sold thereafter.

There may be a certain minimum annual payment required which, if not paid, causes the license to either terminate or, if it is an exclusive license, to convert to a nonexclusive license giving the licensor the right to license others. There may also be a ceiling or upper limit on the payment a licensee would be required to make in any given year and/or during the life of the license agreement. For example, the license agreement might provide that the licensee pay a 10% royalty on the sales price of licensed products sold, but in no event more than $100,000 per year regardless of the number of units sold during the year. In addition to or as an alternative to such a yearly limit on the total royalties, the license may provide that once the total royalties paid in all years have reached $1 million, which may take three, five, or seven years depending on the licensee's sales volume, no further royalties are due during the remainder of the life of the license.

The agreement may also require the licensee to make a down payment of a specified dollar amount in addition to making royalty payments based on production volume. The down payment may be applied against future royalties or may be in addition to the royalties. License agreements sometimes provide that any inventions the licensor makes subsequent to signing the license agreement are automatically included in the license, with or without payment of additional monies.

If a license to a first party is nonexclusive, the license may contain a provision that should the licensor subsequently grant a license to a second person on terms more favorable than those to the first licensee, the more favorable terms are automatically extended to the first licensee. A provision may also be included respecting the right of the licensee to grant sublicenses to others.

Sometimes license agreements contain provisions respecting enforcement of the patent should an unlicensed party infringe it. Such provisions typically specify who shall have the right or the obligation to enforce the patent, if anyone; who shall bear the expenses of enforcing the patent; and who shall retain any monetary recovery if the patent is successfully enforced. A license may contain a provision that if there is substantial infringement by an unlicensed party, the licensee's obligation to continue paying royalties to the licensor is terminated, suspended, or at least the royalty payments placed in escrow pending action by the licensor, resulting in termination of the infringement by the unlicensed party.

In many cases, particularly when exploitation of the patented invention on a commercial scale requires substantial technical know-how and assistance from the licensor, the license will provide for the furnishing of technical information and assistance to the licensee necessary for "start-up." For example, the licensor may be required to provide detailed engineering drawings specifying dimensions, tolerances, materials, finishes, and the like. The licensor may also be required to train the licensee's personnel.

A license granting patent rights *and* know-how has been called a hybrid license. Termination can be set for a period beyond the life of the patent. If so, provision should be made for a reduction in royalties after termination of the patent, the reduced royalties reflecting only the value of the know-how.

Some license agreements impose specific duties on the part of the licensee relative to the degree of effort he must use in exploiting the licensed invention. For example, the licensee may be required to acquire production capabilities of a specific magnitude, spend certain minimum annual amounts for advertising and promotion of the licensed product and the like.

The license may also contain provisions to cover the possibility that the licensee will be charged by another with infringement by reason of manufacturing and selling the licensed invention. Ordinarily, the mere licensing of an invention to another does not carry with it any indemnification of the licensee by the licensor should the licensee, as a consequence of manufacturing and selling the licensed invention, infringe another's patent. However, the parties may provide in the agreement that should an infringement charge be made against the licensee by another, the license terminates; or the royalty will be applied against the cost of defending the licensee; or the licensor and licensee will jointly undertake to defend the suit and share expenses and any recovery; or that the licensor will pay all or certain costs connected with the defense, such as attorneys' fees, damages should the claimant be successful, etc.

If a patent application is involved, the license agreement will often specify which party has the responsibility for prosecuting the application to the point where it becomes a patent, including who is responsible for the cost of obtaining the patent and who shall control the effort to obtain it.

License agreements usually contain some provision with respect to the duration of the license and the manner and the conditions under which it can be terminated. The license may be for a specified number of years subject to periodic renewal or it may be for the life of the patent. One or both parties may have the right to cancel upon giving some predetermined notice, e.g., six months.

The license may also contain provisions to the effect that the agreement can be terminated for cause, such as if the licensee defaults in his payments or the licensor fails to enforce the patent against an unlicensed infringer should he have undertaken the obligation to do so. The license may also contain a provision for arbitration of disputes arising under the agreement. Arbitration is often a prompt and inexpensive alternative to litigation in a court for resolving disputes between the licensor and the licensee.

The license will usually provide that the licensee will, where possible, mark products manufactured under the patent with the patent number. Absent patent marking, damages cannot be obtained from an infringer until after he has actual notice of the patent.

There are also a number of provisions routinely placed in a license agreement such as warranties of ownership of the patent by the licensor, the manner and method of giving notice to the other party where such may be required under the terms of the license, the state whose laws will govern validity and interpretation of the terms of the license, when and under what circumstances the rights of the parties under the agreement can be transferred to others, and whether the agreement contains all the terms that the parties have agreed upon and/or supersedes any specific previous agreements between the parties.

Incidentally, where an invention is owned by more than one person, each co-owner is free to license the patent as if it were his own and retain whatever revenues result. To avoid problems arising from this situation, co-owners of an invention often agree with each other that the invention can only be licensed upon consent of all co-owners, in which event the co-owners share on some mutually agreeable basis all revenues obtained. See Chapter 12.

16

Foreign Patents

Patents are grants from the government, whether it be the Government of the United States as is the case with a United States patent, or the Republic of France in the case of a French patent. Since the some 200 or more governments which comprise the world today are all sovereign nations, their respective laws, and hence the patents granted pursuant thereto, can extend only to, and not beyond their respective geographic boundaries. Thus a patent granted by one country can have no direct legal effect on the activities of a person in another country. Simply stated, the patents granted by a country, like its laws, have no extraterritorial effect.

Considering the United States as illustrative, the grant of a patent on an invention by the U.S. Government enables the owner, whether he be a U.S. citizen or alien, to exclude anyone from making, using, or selling the invention in the United States (and its territories and possessions). The owner of a United States patent can stop others from manufacturing in the United States articles covered by the patent even though the products are contemplated for use and sale abroad. Similarly, the owner of a U.S. patent can stop the sale or use in the United States of articles covered by his patent whether manufactured in the United States or abroad.

However, the holder of a U.S. patent cannot by virtue of that patent prevent a Canadian manufacturer from manufacturing and selling in Canada to Canadian users products patented in the United States nor the export of Canadian-manufactured goods to users in Mexico.

If there is a substantial foreign market for a product patented in the United States and the foreign market is likely to be exploited by a foreign manufacturer, you should consider seeking foreign patent protection. Exactly in which country or countries you seek patent protection will depend, to a large extent, on the location of the particular foreign market and the location of foreign manufacturers capable of satisfying the needs of that market. For example, if there is a market for the product throughout Europe and the product is such that it could likely be manufactured in any one or more European countries for distribution in the country of manufacture as well as in neighboring European countries, then you will have to obtain foreign patents in each European country to preclude sale and use of the product in each European country.

On the other hand, if the market is throughout Europe but by reason of the technology involved the product is likely only to be manufactured in a single country, e.g., Germany, although exported to the rest of Europe, then you can likely preclude exploitation of the market for your product throughout Europe by obtaining and maintaining only a single foreign patent, namely, in West Germany. Although the West German patent will enable you to stop manufacture only in West Germany, this is adequate under the assumed facts since only a West German company has the technology to manufacture the product.

Often the cost of a foreign patent approximates, and in many cases exceeds, the cost of obtaining a patent in the United States. It may not be economically feasible to obtain patents in every country where you may need one to protect the market for your product. If such is the case, you will have to select those countrie s having the biggest market and/or where the most formidable competitors exist and file on a selective basis in only those countries.

Sometimes, your company has a marketing organization in only a certain select number of foreign countries even though there may be a market for your company's product in a number of other countries. In such case, you will typically find that your company will seek foreign patent protection only in those countries where it has an established marketing organization. This makes sense since a company usually obtains foreign patents to avoid foreign competition. For example, consider the situation wherein a market for your product exists in Australia but for one reason or another you have no marketing organization there and do not contemplate having one. In this situation, you would not likely obtain an Australian patent. Sales in Australia of products incorporating your invention, even though by another, are not really "lost sales" for you since you have no marketing organization there to get the sales anyway.

As is apparent from the foregoing, as a general rule a company does not usually obtain foreign patents in countries where it is not itself competing for sales. One principal exception to this general rule, however, is in those situations where a company has substantial prospects for licensing a foreign patent in a particular country even though it does not itself market there. Some companies are very active in the area of patent licensing and generate substantial revenues from it. For such companies, it is natural to obtain a patent in any country where there is substantial licensing potential.

The patent laws of the various foreign countries differ to a large degree. It is not possible within the confines of a single chapter to detail all the similarities and differences. Even if it were, the utility of doing so is questionable. The laws of the various countries undergo continual change. In the final analysis, you will always find it best when considering patenting in a particular foreign country to consult your patent lawyer, who can then focus on the intricacies of the patent laws of that country and advise you. Without getting into specifics, there are, however, some generalizations to be made about foreign patenting which will give you insight into the kinds of problems that arise and with respect to which decisions must be made.

As you learned in Chapter 3, in the United States it is not possible to obtain a valid patent if, more than one year before the effective filing date of the United States patent application, the invention was described in a printed publication published anywhere in the world or the invention was in nonexperimental public use in the United States. The same is true in Canada except that there is a two-year grace period permitted within which to file rather than one year as in the United States.

Some countries are more lenient than those discussed above while others are stricter. Illustrative of countries where the rules are stricter are France, Italy, West Germany, Great Britain, Israel, Mexico, and the Scandinavian countries. In these countries a valid patent cannot be obtained if, any time before the effective filing date of a foreign patent, the invention was described in a printed publication available anywhere in the world or was known or used anywhere in the world. These latter countries and others like them are known as "absolute novelty" countries. Thus, unlike the United States and in France, Italy, West Germany, Great Britain, Israel, Mexico, and the Scandinavian countries, for example, knowledge or use of the invention anywhere in the world, any time, even one day prior to the effective filing date of the patent, will bar issuance of a valid patent.

In view of the one-year grace period provided by the United States patent laws, a company often will first test-market a product for six or eight months and, if it proves a commercial success, seek a U.S. patent. As long as the U.S. patent application is filed within one year of the first non-experimental commercial use of the invention which, to be on the safe side, will usually be considered the first offer for sale (even though the product may not actually have been sold) or public use or public disclosure, a valid U.S. patent results. However, since the absolute novelty countries such as France, Italy, Mexico, West Germany, Great Britain, Israel, and the Scandinavian countries require that the effective filing date of the patent be prior to public disclosure of the invention anywhere in the world (vis-a-vis, within one year following public disclosure), test marketing in the United States will bar obtaining a valid patent in these strict novelty countries if it is done prior to the filing date of the U.S. patent.

Thus, if you contemplate obtaining valid patent protection in any of the strict novelty countries in addition to protection in the United States, you must file the U.S. patent application prior to any public use, disclosure, or sale of the invention in the United States. Of course, if you do not contemplate obtaining patent protection in a strict novelty country but rather only in countries having laws patterned after the United States where there is a grace period within which to file for a patent after public use, e.g., Canada, or in countries more lenient than the United States in this regard, then you can test-market your product in the United States and file for a U.S. patent within the one-year grace period if the product proves successful. The grace period is a changing situation. Consult your patent lawyer for updated law.

Once an initial selection is made of those countries where foreign patents will be sought, based on marketing and/or licensing considerations abroad, there are other factors in addition to cost to consider which may cause you to further limit the number of countries where you seek patents. For example, in some countries, the degree of unobviousness or quantum of advance in technology needed for grant of a patent is exceedingly high, whereas in other countries the standard by which patentable inventions are measured is relatively low.

It is interesting to note that in those countries requiring relatively high degree of inventive merit for the grant of a patent, e.g., West Germany, when a patent is obtained and subsequently asserted against an infringer, the validity of the patent is typically upheld by the court and accorded a fairly broad scope. In those countries in which patents are relatively easily obtained, a larger percentage of patents are invalidated by the courts or the scope of protection restricted.

Perhaps the single most important factor in deciding in which countries foreign patent applications will be filed is the expense involved. One might think that once a patent application has been drafted and prepared for filing in one country, the cost of obtaining and maintaining foreign patent protection would be relatively modest, involving little more than a "routine" duplication of effort abroad. Unfortunately, this is not the case. In analyzing the cost of foreign patents, there are a number of factors to consider. First, if protection is sought in a non-English-speaking country, a translation of the patent application into the language of the foreign country of interest will be required. This is a significant expense, particularly for a patent application which has a lengthy specification.

In addition, in most countries the claims and, in many cases, the introductory portion of the patent application will have to be reworked to place them in a form consistent with the requirements of the different foreign countries, which often is quite different from that of the United States.

If the country is one where examination of the invention with respect to novelty is required, i.e., an examination is made to determine if the invention is unobvious over the prior art, as in the United States, significant expense can be expected while the application is pending. The foreign Patent Examiner will cite prior patents and publications. These will have to be studied and arguments presented to distinguish the invention over them. In addition, it may be necessary to amend the claims to distinguish over the prior technology, particularly where it is a foreign patent which may not have been cited during the pendency of the corresponding United States application. These activities cost money. On the other hand, in those countries where the invention is not examined with respect to prior technology but only with respect to formalities such as whether the claimed invention falls within one of the categories of inventions for which patents are granted, whether a single invention is claimed or a plurality of inventions, etc., there will usually be only minimal expense during pendency of the patent application.

Finally, in the vast majority of foreign countries annual taxes or maintenance fees must be paid to avoid lapse of the patent. Typically, the annual fees increase in size each year; in some countries reaching as high as several thousand dollars the last year. As a matter of interest, the theory behind renewal or maintenance fees is that patents on inventions which are of no commercial interest to the owner will be permitted to lapse and the subject matter thereof fall into the public domain available for noninfringing use by all. Further justification of renewal fees, particularly in examination countries, is that the cost to the government of granting a patent is greater than the fees paid during the examination phase and this excess cost should be borne in the form of annuity fees by those who benefit from the system, namely, by those who obtain patents.

Thus, when considering the cost of patenting abroad, one cannot consider only the cost of filing the foreign application, such as government filing fees, translation costs, and reworking the introduction and claims of the patent application to place them in compliance with foreign patent practice, but also the cost of doing battle with the foreign Patent Examiner in those countries where the invention is subjected to an examination for novelty prior to granting the patent and the escalating annual renewal or maintenance fees which must be paid throughout the life of the patent to prevent it from lapsing.

Although perhaps not influencing the decision as to where to seek patent protection, there are a number of other differences in the patent laws between the United States and foreign countries of which you should be aware. For example, in the United States the duration of a patent is measured from the date it is granted, whereas in some foreign countries the life of a patent is measured from the date the patent application was filed, which may be anywhere from one to three years or more prior to the patent issue or grant date. In many foreign countries, the duration of the patent is not 17 years, as the United States, but may be a lesser period, e.g., 15 years, or a greater period, e.g., 20 years.

Some countries automatically publish a patent application within a predetermined time after filing, e.g., 18 months from the filing date of either the foreign application or its U.S. counterpart. In some situations where obtaining a patent is problematical due to questionable merit of the invention, it may be to your advantage to forego foreign filing and the attendant publication, and instead maintain the invention as a trade secret. In this way, although you do not get a patent, you also do not end up publishing it.

Not all countries grant patents on all inventions. For example, the Scandinavian countries do not grant patents on drugs. By way of contrast, in the United States pharmaceutical patents are routinely granted and have proven to be extremely valuable in many cases.

In some countries which require examination as to novelty before granting a patent, the novelty examination is not automatic but rather only done at the request of the applicant. In such countries, the expense of the novelty examination and the preparation of arguments and amendments to the claims to distinguish over cited prior technology may be deferred until such time as it appears that the actual grant of a patent in that country is absolutely necessary. In other countries, which provide for a novelty examination before the grant of a patent, the novelty examination is automatic and cannot be deferred, as is the case in the United States.

Unlike the United States, a number of foreign countries have what is known as *opposition proceedings*, whereby a patent application, after it has been examined by the Patent Office and found to be worthy of issuance, is published for a specified period, e.g., three months, within which time any interested party can oppose issuance of the application, e.g., on the ground of obviousness over a prior patent not considered by the Examiner. The Scandinavian countries and West Germany are illustrative of only a few of the many countries providing for opposition. In Great Britain there is a proceeding after issuance of a patent, termed a *revocation proceeding*, which in many respects is like an opposition.

Another contrast of interest between the U.S. patent system and that of many foreign countries is that in the United States if different parties on different dates file patent applications for the same invention, an *interference proceeding* is instituted to determine which party made the invention first. Thus, in the United States the person who ultimately gets the patent (where two or more persons filed application for the same invention) is the one who proves he made the invention first, whereas most foreign countries do not have an interference proceeding or anything equivalent to it but rather decide who gets the patent on the basis of which party filed his application first. Thus in most foreign countries the party who has the earlier effective filing date is the party who gets the patent.

Some countries also have what is known as a *working requirement*. In a country having such a requirement, unless a patented invention is commercialized in the country granting the patent within a specified period from grant, e.g., three years, the patent is subject to revocation or compulsory licensing at a scheduled royalty to anyone who applies to that country's Patent Office for a license. The United States has no such requirement or anything comparable to it.

In a rather limited number of countries, of which Germany and Japan are the most industrialized, *petty patents* are available. A petty patent is inferior to a conventional patent in terms of the duration of the patent and the types of inventions which can be covered. There is, however, a corresponding reduction, when compared to a conventional patent, of the degree of inventive merit required to obtain a petty patent.

In a number of countries, Canada and the United States being notable exceptions, a patent owner may obtain a *patent of addition* on a modification of his basic patented invention, which modification does not itself represent a sufficiently unobvious advance to justify grant of a regular patent. The term of a patent of addition is usually limited to the unexpired term of the principal patent although there are some exceptions.

In approximately 25 countries, most of which are Central or South American, *patents of importation* may be obtained. In effect, a patent of importation is the registration in that country of the first patent obtained elsewhere on the invention and usually can be obtained if the invention described in the patent of importation had not previously been publicly used in the country where the patent of importation is sought. Publication of the invention, such as by the grant of the first patent, will not act as a bar to a patent of importation. As a general rule, the duration of a patent of importation will be coextensive with the unexpired term of the foreign patent upon which it is based.

No discussion of foreign patents would be complete without mention of present and prospective treaties which exist among various nations which bear on patenting abroad. Perhaps the oldest and most well known is the Paris Convention of 1883, revised numerous times with the latest being in Stockholm in 1967. By the terms of the Paris Convention, a second application filed in a "convention country," i.e., a country which ratified and thereby became a member of the convention, within 12 months after the filing of the first application in any other convention country has the same force and effect in the second country as if it had been filed simultaneously with the first-filed application. The practical effect is to give an applicant who has filed in his home country (or any other country) one year within which to file applications in other countries. Of course, the one-year grace period provided by the Paris Convention will not preclude barring of protection in the strict novelty countries if, prior to the filing of the first application, the invention was publicly known or published anywhere in the world.

The African and Malagasy Union, to which over 10 African countries, all previously members of the French community, have become members, is the first instance anywhere in the world where separate and independent countries have created a patent law which facilitates the grant of a single multinational patent covering all member countries. There are also other treaties in various stages of negotiation and ratification directed to international cooperation in filing and granting other forms of multinational patents.

The Patent Cooperation Treaty (PCT) purports to simplify filing of patent applications in many member countries. In the PCT, a single patent application is filed and searched in a single designated office, such as the United States Receiving Office located in our own U.S. Patent and Trademark Office in Arlington, Virginia. Thereafter, the application and PCT search report are forwarded to each designated country for national patent proceedings.

Alternatively, under new procedures, referred to as PCT-Chapter II procedures, a single examination to determine patentability can also be conducted and the patentability report also sent to each designated country. This can eliminate major prosecution expenses in each country. It is hoped, but not yet fully established, that each member country will simply accept the examination and patentability report and proceed to issue, without further question, patents on those applications favorably examined.

The final result under the PCT procedures is the issuance of a number of different "national" patents, one issuing from each designated country based on the original single application. All generally gain the benefit of the U.S. filing date, so that intervening disclosures of the invention do not invalidate the foreign patent, even though they were later filed.

Under the PCT, certain fees and costs to be incurred for each country can be deferred. This permits an early filing, but with a deferment of costs for a time period perhaps sufficient to analyze the commercial performance of the invention before committing significant patent costs to a particular country. Thus, filing for foreign protection in all PCT countries at minimal cost can be accomplished at an early time before any nonconfidential disclosures of the invention (which would invalidate a patent) while delaying commitment of significant fees until a later time. The opportunity to get a search report, and perhaps a patentability report, on the claims before the designation and other fees are due can help in determining whether or not to commit the fees to proceed in the various countries. Where early issuance of a patent is not a major factor, this proceeding results in both cost deferral and ultimate savings as compared to the old system of a national filing for each foreign country.

The new PCT-Chapter II procedures permit patent owners to more closely and more flexibly manage their patent rights and properties in the PCT member countries than ever before. Programs of protection can be more economically tailored for each invention, and significant fees and expenses delayed and uncommitted for time periods up to some thirty months, adequate in many cases to permit commercial tests of the product market prior to commitment. Consolidation of an international search report, patentability report and prosecution expenses for PCT members can save substantial amounts compared to past procedures. In the final analysis, each case should be managed on an individual basis so that the patent owner can secure the most benefit from its intellectual property dollar. The flexibility and cost savings provided by the new PCT procedures will be very helpful in this effort where multiple countries are involved.

The European Patent Convention (EPC) of 1973, signed by over 11 European nations and effective starting in 1978, provides a single procedure for the grant of patents enforceable in selected designated member countries. An applicant can request that a "European convention patent" be granted for one or more specifically designated member countries upon payment of the appropriate fee.

Once granted, a European convention patent will have the same effect in a designated country and will be subject to the same conditions as a national patent granted by that country.

In the EPC, costs are generally incurred more quickly than in PCT. These are offset by the later savings in prosecution, since only one application is prosecuted to mature into a single patent covering numerous countries.

The EPC and the PCT recognize the existence of each other. There is the possibility of combining their benefits so as to obtain patent protection in each country on your list (non-PCT countries excluded) even though a country may not be a member of both. This permits taking advantage of certain cost savings and fee deferments of the PCT, even for the EPC member countries.

Both the PCT and EPC provide certain filing economies for patent applications directed to their member countries, although their procedures are somewhat different. The procedures for both the PCT and EPC are complex and have many variations. A review of the scope of domestic and international protection should ultimately be made concurrently with the disclosure of each new invention to patent counsel so that a plan can be programmed most efficiently for each case.

It still remains to be seen whether patents covering or designating multiple countries under these conventions and treaties will really constitute overall economic, commercial, or legal advantages or disadvantages over separate national patents.

Again, and in view of the varying laws of each country, an inventor or company must keep in mind that *where foreign protection is desired, it is critical to file the U.S. (or PCT patent application) prior to any non-confidential disclosure of the invention. If the invention is disclosed non-confidentially in any manner prior to an appropriate patent application filing, the right to obtain any valid foreign patent in most countries will be lost.*

17

Misuse of Patents

O.K., so now you have obtained your patent, know how it can be enforced, and know what types of licensing agreements are available. The time has come, you feel, to enter into a license agreement in which you will require your licensee to purchase only certain unpatented, staple components of the licensed structure from you. Obviously, a sound economic policy, right? Or, as a condition of the license you extract a promise from the licensee that he will not deal in certain items not covered by your patent which your competitor sells. Again, sound economic policy putting increased profits on the balance sheet, right? Economically attractive, *but almost certainly legally impermissible*, upon penalty of having your patent rendered unenforceable, because of the judicial doctrine called "patent misuse."

The doctrine of patent misuse can be traced back to a U.S. Supreme Court decision in the early 1900s. In essence, it is any conduct on the part of the patent owner to expand his patent beyond the bounds provided by law, i.e., to cover unpatented items, to extend beyond 17 years, etc. The misuse doctrine has expanded greatly since that time, but at the present time there are indications that it is being judicially restricted, and there is also a definite possibility that it will be restricted by statute to those situations where the misuse rises to the level of an antitrust violation. Patent misuse, if found, renders the patent owner's patent unenforceable. What this means is that the patent will not be enforced by a court against an infringer. Patent misuse does not invalidate the patent–it just makes it unenforceable. This is an important distinction because a patent owner who has been guilty of misuse can enforce his patent after he has purged himself of the misuse. What constitutes a purge of a misuse is based on the particular facts. It may require the patent owner to terminate a license agreement or abandon an offensive practice.

The effects of a misuse may be worse than a mere finding by a court of unenforceability because some misuses may give rise to an antitrust claim. If proven, such an action can result in a damage award measured by the party's actual damages trebled plus attorneys' fees. Not every misuse constitutes an antitrust violation.

A misuse which is not an antitrust violation can also create monetary liability in the form of an award of attorneys' fees to the defendant who proves such if the court is convinced that the case is an "exceptional" one and attorneys' fees should be awarded. Although such an award in an infringement case is infrequent, it nevertheless is a possibility that must be considered if a patent owner has been guilty of questionable conduct in the enforcement or licensing of his patent.

Generally speaking, any attempt to procure, under the guise of a patent license or through other conduct, a monopoly or the exclusion of competition through the use of one's patent, and which is not justified under the patent laws, can be urged as a patent misuse. It is almost a certainty that any questionable conduct will be urged by an accused infringer as a misuse. With such a broad spectrum of conduct prohibited, it might be well to first look at what the owner is legally entitled to do before looking at specific courses of conduct which have been held to constitute patent misuse.

The patent laws specifically provide for granting a territorial license under the patent; hence, such can be entered into. An exclusive license can likewise be granted. A suit for patent infringement against one who is making, using, or selling the invention in the United States can be brought if reasonable grounds are present for believing there is an infringement.

So much for the clear-cut permissible areas of licensing and conduct. Now, let us turn to the nonpermissible and gray areas. A word of caution at this time; there are many gray areas which can become impermissible depending upon other conduct of the patent owner. The particular conduct by itself may be proper but when it is combined with other conduct, likewise permissible by itself, the total course of conduct may constitute a patent misuse. So we do not look at a specific possible course of action in isolation, but rather in the total context of the business in question and the effect of the total program on your competitors.

The classic example of patent misuse is exemplified in the situation where the patent in question covers (1) a chemical process which utilizes staple unpatented chemicals, or an apparatus or article which is made up of unpatented elements, and (2) the patent owner will only license the patent on the condition that the licensee buy the unpatented chemicals or elements from him. This constitutes patent misuse—a "tie in."

The requirement to purchase the staple from the licensor may stem from an actual provision in a license agreement to do so or it may be implied from the patent owner's conduct. No agreement setting forth this purchasing requirement need be entered into or proposed if the net result is to force the licensee to buy only from you. For example, a more subtle route that has been unsuccessfully tried in the past is to grant a license on terms such that the licensee pays a unit patent license royalty of X if he buys the unpatented component from the licensor and 2X if he buys from the licensor's competitors. Again, the net effect is the same; it forces the licensee, or at least encourages him, to purchase the unpatented staple component from the licensor. What is improper about this is that the conduct is outside of the rights granted to the patent owner.

A substantial number of "tie-in" misuse cases do arise. In the classic example given above, the patent owner's business usually is selling the unpatented staple compound and he does not himself practice the patented process. Most often there are other competitors who have been selling the unpatented staple for other uses before the patent issued. While the patent owner would like to be the sole supplier for the unpatented staple compound for use in his patented process, his patent right only extends to the right to exclude others from using the *process* (vis-a-vis the *unpatented compound*). Nevertheless, the patent owner will license his process patent, often without change, providing the licensee buys the unpatented staple compound from him, creating a tie-in misuse.

The safest course of action to avoid a tie-in misuse is to license the end user of the staple in a form which would permit him to purchase the unpatented compound, without monetary penalty, from any source. In many instances this can be accomplished through a device known as a *label license* whereby a notice on the compound's container advises the purchaser that the purchase price includes a prepaid royalty of X cents for a license to use the patented process which, upon request, will be refunded if the patented process is not used and that the purchaser can use other manufacturer's compounds by paying the same X cents royalty.

The foregoing discussion demonstrates one of the reasons that a process or use patent is less desirable than a product or composition patent under which one obtains the right to prevent others from selling the patented composition.

We have just been talking about the tie-in of unpatented, *staple* products. These may be defined as being products which have a number of uses in addition to use in the patented process. But what of a product which has no other use than in the patented process and is a material part thereof?

The patent statute gives to the owner of a patented combination or process the right to prevent others from selling a material component used in the patented combination or process if the component is not a staple. The Supreme Court has interpreted this to mean that the owner of a process patent which utilizes a nonstaple chemical product in the practice thereof may sell the product and grant with it a license to use the product in the patented process, and at the same time refuse others the right to sell the product for the patented use and refuse to the users of the process the right to buy the product elsewhere.

What, then, are the other areas of *potential* patent misuse? Although certainly not all-inclusive, the list of most frequently encountered areas includes license agreements or conduct in which the licensor

1. Fixes prices
2. Establishes exclusive dealership
3. Establishes discriminatory royalty rates between competing licensees

4. Extends the payment of royalties under the license beyond the expiration date of the patent
5. Requires the licensee to take a license under a package of patents in addition to the one the licensee desired

We have referred to the above areas as potential misuses because not each one will always constitute a patent misuse. For example, although the U.S. Justice Department has fought vigorously to have the decision overruled, there still remains a U.S. Supreme Court decision which *may* still permit a nonmanufacturing licensor to set the sale price at which his only manufacturing licensee may sell.

The requirement imposed on a licensee to deal only in the goods of the licensor as a condition of obtaining the license is a frequently found patent misuse. Here again such a requirement is not a right granted under the patent laws.

Normally a patent owner has the right to set whatever royalty rates he chooses. However, misuse problems can be encountered, as several courts have found, if he licenses two licensees at different rates and if the different rates have the effect of eliminating or injuring competition between the licensees. Several recent cases refused to rule that such was a misuse when it was determined that there was a valid economic justification for the different royalty rates, or the licensees were not in direct competition, or the royalties were not an important cost factor.

Quite naturally, a licensor desires to receive royalties for the licensed patent property for as long as possible. The patent laws provide that a patent expires in 17 years, at which time anyone can practice the invention without payment to the patentee. A licensor who enters into a license agreement whereby royalties are payable beyond the 17-year period is attempting to reap more than he is entitled to and will probably be guilty of misuse. The application of this doctrine can be especially troublesome in a situation where multiple patents having different expiration dates are licensed at a fixed royalty which extends to the last-to-expire patent. Misuse has been found where the "package" of patents included one which, by comparison with the others in the package, was dominant and very important and no diminution of the royalty rate occurred upon its expiration. Such an arrangement was held to be tantamount to the collection of royalties on an expired patent.

Closely akin to the immediately preceding situation is one in which the licensor requires the licensee to take a license under a package of patents in order to obtain a license under the patent(s) he desires. Such conduct has been held to be a patent misuse. A licensor should instead be prepared to negotiate in good faith for a license under the desired patent(s).

As pointed out at the outset, misuse can be purged by eliminating the objectionable conduct. The purge must also include the elimination of the effects of the misuse. This latter requirement may mean that even though the offensive conduct has been eliminated, enforcement of the patent will be delayed until the undesirable effects of the misuse have been dissipated, i.e., until such time as the original status quo, i.e., prior to the misuse, has been achieved.

18

Trade Secrets

For about 100 years, the courts of the United States have accorded a trade secret owner the right to prevent disclosure or unauthorized use of the trade secret by those who acquire it by improper and unfair means or who seek to convert it to their own use in breach of a confidential relationship. This right was reaffirmed by our U.S. Supreme Court in 1974. In essence, the right protected is one which stems from an obligation of fairness, and the cornerstone of the right is that one's private business information which gives him a competitive advantage will not be taken from him wrongfully.

The protection of trade secrets extends to many different types and forms of technical and/or business information. A trade secret may be a chemical formula, a chemical process, a pattern for a machine, drawings and blueprints, tolerance information, customer lists, a prototype, model, photograph, financial information, supplier information, and the like. In other words, it is any type of confidential information used in one's business which gives him a competitive advantage over his competitors.

In order to be protected as a trade secret, there must be some degree of secrecy with respect to the claimed trade secret, and it must be shown, as discussed below, that the trade secret was or is about to be misappropriated by unfair or illegal means. The degree of secrecy required has been considered by many courts over the years and there has evolved no single rule quantifying the degree of secrecy required. Situations ranging from disclosure in a Ph.D. thesis to general knowledge in the industry have been involved.

In seeking to establish trade secret rights, it is important that the trade secret owner be able to point to steps he has taken to preserve the secrecy of the trade secret. For example, if the trade secret is a chemical formula, it would be pertinent for the owner to show how many formula books are in existence, the people to whom they have been given, and the steps taken to preserve their secrecy. Employment contracts with key employees wherein the employees agree not to divulge the trade secret are relevant. So, too, are steps to prohibit unauthorized visitors to the plant or other location where the trade secret is being practiced. It is easier to convince a court that blueprints which contain a confidential legend on their face are in fact confidential than ones which do not.

Misappropriation of Trade Secrets

A trade secret's life is similar to that of a balloon at a child's birthday party. It can disappear just as quickly because trade secrets are protected only as long as they are not known in the industry. What this means, for example, is that if one of your competitors is marketing a new dishwashing detergent, you are free to "reverse-engineer" it, i.e., to analyze the product and to market an identical product even if your competitor claims trade secret rights in the formula. This assumes, of course, that he does not have a patent covering it. This is true because you have done nothing which our law regards as unfair. However, if you choose to accomplish the same result, but instead of analyzing the product you hire away one of his employees who has knowledge of the formulation, and you obtain the information from that employee in violation of his duty to his former employer to keep it a secret, you will be liable for the misappropriation of the trade secret, assuming, of course, the presence of the other required elements, secrecy and unfair or unlawful misappropriation, discussed herein.

This last situation is the one which most frequently arises because employees often change jobs. The courts, even in the absence of a written employment agreement requiring secrecy, imply a relationship between the employer and employee of confidentiality and require that the employee not use or disclose his employer's trade secrets either during employment or after the employment relationship has ended. In several states, courts have permitted an employee who himself has developed the trade secret, presumably for his employer, to use the trade secret after he leaves the job.

Many employers today require their employees to sign a statement agreeing not to use or disclose the employer's trade secrets after the employment has terminated. Such agreements, if fair in other respects, are enforceable and an employer can obtain an injunction against the use or threatened use by the ex-employee of the trade secret. Such agreements frequently cover not only trade secret information but information of lesser importance that is still confidential. Such agreements, too, are normally enforceable. Such agreements must be carefully considered before new employment is sought since a highly specialized employee may find his employment possibilities substantially limited by reason of his specialized knowledge.

As mentioned earlier, the protection of trade secrets stems from an obligation of a business competitor not to use unfair and improper means to learn the trade secret. As we have seen, this precludes hiring ex-employees in order to learn the trade secret. It would be unfair and improper also to enter into license negotiations with the trade secret owner with an avowed purpose of evaluating the information, and then to not take a license but instead appropriate the trade secret without the owner's permission. It has been held by one court that improper means were used to learn of a trade secret when someone hired an aerial photographer to fly over his competitor's chemical plant while it was under construction to take pictures so as to learn more about the competitor's chemical process. Obviously, it would also be improper to break into a competitor's plant for the purpose of learning his trade secrets.

The employer-employee relationship is not the only relationship in which an implied obligation not to use or disclose another's trade secret may be found. Such a duty has been found between

1. Manufacturer and independent contractor
2. Supplier and purchaser

3. Corporate directors, who are not employees, and the corporation
4. Partners
5. Joint ventures

Finally, many states have adopted criminal statutes which make it a crime to misappropriate trade secrets. Such statutes have been enforced and people successfully prosecuted thereunder. In one situation an employee of a large soap and toilet goods company was successfully prosecuted for attempting to sell an advertising program to his employer's competitor. There is also a federal criminal statute which makes it a crime to transport stolen property across state lines. Successful criminal prosecution has been achieved under this statute for transporting stolen trade secrets across state lines. In one situation employees of a large oil company were prosecuted for transporting across state lines maps which showed the possible location of valuable oil reserves.

Trade Secrets Contrasted to Patents

Unlike a patent which expires in 17 years, a trade secret owner's rights extend for as long as the trade secret is not known in the industry. Also unlike a patent, anyone may use a trade secret as long as it is obtained fairly. Formulas which are protected and which cannot be determined through analytical techniques may be protected for many years. Illustrative of a trade secret which has endured for many years is the formulation for Coca-Cola brand soft drink.

Unlike a patent, there is no requirement that a trade secret display that amount of invention which would make it patentable. In other words, a trade secret does not have to be novel and unobvious to one of ordinary skill in the art. It may be only a minor step forward in the art or it may be no step at all. It can be a customer list which a competitor could not come up with on his own without a great deal of time and expense. It may be mechanical tolerances which could not, without the expenditure of a great deal of time and money, be arrived at by one's competitor.

The Decision to Keep the Secret or to Patent It

When something new is developed which could be patented, a determination should be made at some point as to whether it should be protected as a trade secret or patent protection should be sought. For example, assume one develops a new process for making a chemical product and there is no way of determining the process from the product itself. Assume, too, that patent protection could be obtained. The option is then open to protect the development as a trade secret or to patent it. The decision can be postponed until the U.S. Patent and Trademark Office acts on the patent application since the proceedings are held in secrecy and no destruction of the trade secret occurs until the patent issues. Some of the important considerations would be

1. What are the chances that some other company will itself discover the process through independent effort?
2. What are the probabilities that secrecy can be maintained?
3. Is it likely that if the patent issues others will use the invention undetected?
4. Is it likely that if the patent issues others can "design around" the patent?

Another consideration is the likelihood of someone else discovering the secret and then patenting it. This could present patent infringement problems since use of a process in secret will normally not be a defense against infringement of another's patent.

Exploitation of Trade Secrets

Just as in the case of patents, trade secrets may be licensed and royalties for the use thereof collected. Trade secret license agreements normally include many of the provisions that one finds in patent license agreements. Special emphasis is usually placed on the requirement of the licensee maintaining the confidentiality of the trade secrets because, obviously, if the trade secret becomes public knowledge it also loses its status as a trade secret and its value is seriously diminished. Frequently, patent license agreements also include, as an adjunct, a technical know-how license which may be, in reality, a trade secret license.

19

Copyrights

As of January 1, 1978, copyrights are controlled exclusively by federal (vis-a-vis state) law. Prior to January 1, 1978, the field of copyright law was governed by both a federal statute, enacted in 1909, and laws of the individual states. The discussion which follows describes the law as it is now, i.e., the new law.

A copyright is a grant of certain rights by the U.S. Government to the originator of a copyrightable work. If the work was commissioned, i.e., the originator did it for hire, the copyright is granted to the one who commissioned it. When contracting with any nonemployee to create a copyrightable work, the contract should be in writing and state that the client owns the work and all copyright rights therein.

Works of the type which are copyrightable include

1. Literary works, e.g., books, periodicals or contributions thereto, cartoons, comic strips, advertisements published in newspapers and magazines, radio and television scripts, letters, diaries, and similar personal manuscripts, poems, and lectures.

2. Music and accompanying lyrics.
3. Dramatic works and accompanying music.
4. Pictorial, graphic, and sculptural works, such as maps, blueprints, works of art or models or designs or reproductions of a work of art, architectural and engineering drawings, plastic work of a scientific or technical nature, prints, photographs, pictorial illustrations, game boards, and labels for an article of merchandise.
5. Motion pictures, including the sound track and other audio-visual works, and sound recordings.
6. Pantomimes and choreographic works.
7. Computer software.

The rights granted to the copyright owner include the right to

1. Reproduce and distribute the copyrighted work.
2. Prepare derivative works such as abridgements, condensations, translations, and the like which are based on the copyrighted work.
3. Perform publicly those works which can be performed, such as literary, musical, dramatic, and choreographic works, pantomimes, motion pictures, and other audio-visual works.
4. Display publicly those works which can be so displayed, such as, pictorial, graphic, and sculptural works.

Each of the foregoing rights can be separately owned and enforced. And like other forms of intellectual property, these rights can be licensed and transferred to others.

The protection afforded by a copyright extends only to the manner in which the copyrighted idea is expressed and not to the idea itself. For example, the owner of a set of copyrighted blueprints for a building, machine, or other product cannot preclude one from constructing the building, machine, etc. He can, of course, prevent others from copying and distributing the blueprints. Similarly, publication in a copyrighted article of a detailed description of an electrical circuit, including its function, use, method of construction, and the like, does not give the owner of the copyrighted article the right to prevent someone from manufacturing, using, or selling the circuit. If protection of the latter type is desired, a patent must be obtained, assuming, of course, that the requirements for a patent—namely, novelty, utility, and unobviousness—can be met.

Certain things cannot be copyrighted. Among these are abstract ideas, procedures, processes, systems, methods of operation, concepts, and principles. This is true regardless of the manner in which the idea, procedure, etc., is expressed, whether it be by words, illustration, or in some three-dimensional form. Of course, the manner in which the idea, procedure, etc., is expressed can be copyrighted.

A copyright on the same work, e.g., a drawing of a machine part, can be granted to multiple individuals providing each originated it independently of the other, i.e., one did not copy it from the other. It makes no difference who created the work first, each is entitled to a copyright as long as each created the work independently of the other.

This is in contrast to a patent which can be granted only to the person who made the invention. Thus, if two people make the same invention independently of each other, only the one who can prove he made the invention first gets the patent.

The copyright owner is given the exclusive right to make and distribute reproductions of the copyrighted work. To a limited extent, however, others can do so and not be infringers of the owner's copyright. This is permissible under what is known as the "doctrine of fair use." For example, under certain conditions, libraries and archives can make a reproduction of a copyrighted work without infringing. Similarly, teachers can make reproductions for classroom use and critics can quote from copyrighted works in reviews.

In considering whether a reproduction of a copyrighted work is "fair use" and not an infringement, the following factors are considered:

1. The purpose and character of the use, including whether it is commercial or for nonprofit educational purposes.
2. The nature of the copyrighted work.
3. The amount and substantiality of the portion used when compared to the copyrighted work as a whole.
4. The effect of the use on the potential market for, or the value of, the copyrighted work.

While perhaps an oversimplification, copying of another's copyrighted work will be permissible under the doctrine of fair use if it does not unreasonably deprive the owner of profits he might otherwise obtain but for the copying.

Under the current copyright law, which became effective in January, 1978, for most copyrighted works copyright protection begins when the author creates the work and extends for the life of the author and for an additional 50 years after the author's death. Under the old copyright law, works had a term of copyright of 28 years with the right to renew for an additional 28-year period. Thus, a total period protection of 56 years was possible. To accommodate the transition between the old and new laws, a copyrighted work which was published prior to January 1, 1978, and which had not expired under the old law prior to January 1, 1978, will have, pursuant to the new copyright law, a 75-year term from the date the copyrighted work was first published, irrespective of when the author dies. Of course, if the copyright had expired under the old law prior to January 1, 1978, it remains expired and cannot be renewed. Thus any work whose copyright expired under the old law prior to January 1, 1978, is in the public domain available for use by anyone.

To insure that a copyright is obtained, whenever copies of a copyrighted work are distributed a "copyright notice" should be placed on each copy in a manner and location calculated to put people on reasonable notice of the fact that the work is copyrighted. Where the copies are visually perceptible, e.g., book, the "copyright notice" must contain each of the following:

1. The symbol ©, the word copyright, or the abbreviation Copr.
2. The year of first publication of the work.
3. The name of the copyright owner, which is usually the originator unless, as noted previously, the work was done for hire in which event the copyright owner is the one who commissioned it.

For visually perceptible works, the copyright notice preferably should be placed on the title page, if there is one, or on the first or front page if the work has no title page. If the work is not visually perceptible but is in the form of a phono record of a sound recording, e.g., a tape recording, phonograph record, etc., the copyright notice must include the following:

1. The symbol P.
2. The year of first publication of the sound recording.
3. The name of the owner of the copyright.

The notice preferably should be placed on the label if a phonograph record or on the cassette, cartridge, or reel if a tape.

Distribution of visually perceptible copies and/or phono records of sound recordings without the proper copyright notice will result in loss of the copyright, except in certain limited circumstances. Accordingly, care should be taken to insure that the copyright notice is provided on all copies. Where visually perceptible copies and/or phono records of sound recordings have been published without the requisite copyright notice, copyright protection will not be lost if one of the following three conditions can be met:

1. Notice is omitted on only a relatively small number of publicly distributed copies.
2. The owner's claim to the copyright has been registered with the U.S. Copyright Office before publication without the notice or within five years thereafter, and reasonable efforts are made to add the notice to copies or phono records publicly distributed after it is discovered that the copyright notice was omitted.
3. The notice was omitted in violation of an expressly written provision which constituted a condition of the copyright owner's authorization of public distribution of copies of phono records.

As noted, for most works copyright protection begins when the work is created and is lost if distributed without the copyright notice if no corrective measures are taken. While not necessary to obtaining a copyright, provision in the copyright law is made for registering the copyright with the U.S. Copyright Office once obtained. There are certain advantages of doing so. For example, registration is necessary before an infringer can be sued. In addition, certain of the remedies for infringement, discussed hereafter, are unavailable where the registration requirements have not been met within a predetermined period of time. Finally, where visually perceptible copies and/or phono records of a copyrighted work were distributed without the requisite copyright notice, registration prior to publication without notice or within five years thereafter is necessary to prevent forfeiture of the copyright.

To be on the safe side, whenever a work is created which falls within one of the categories for which copyright protection exists, a copyright notice should be placed in some conspicuous place on every visually perceptible copy and/or phono record or sound recording. In this way, you can avoid the risk of possible forfeiture of the copyright by subsequent public distribution of visually perceptible copies and/or phono records without the copyright notice.

Where infringement of a copyright occurs, remedies are available including impounding and destroying the infringing articles, actual damages suffered as a result of the infringement, profits of the infringer attributable to the infringement providing they are not accounted for in computing actual damages of the copyright owner, an injunction against continued infringement, court costs, and attorney's fees. In those cases in which there has been no actual damage to the copyright owner or profits by the infringer, the copyright owner may elect to recover damages in accordance with a statutory schedule of not less than $250 and not more than $10,000, except where the infringement was willful, in which case the ceiling may be increased to $50,000. The lower limit on damages may be reduced to as low as $100 where the infringement was innocent. A civil suit for copyright infringement must be initiated within three years after the copyright infringement claim arose. If it is not, the civil remedy for copyright infringement is lost.

There are also criminal penalties provided for willful infringers, including a fine of not more than $10,000 or imprisonment for not more than one year, or both. However, where the infringement is of a sound recording or motion picture, the infringer may be fined up to $25,000 or imprisoned for one year, or both, for the first offense; and fined not more than $50,000 or given two years imprisonment, or both, for subsequent offenses. A fine of $2,500 may be levied against one who places a fradulent copyright notice on a work, or removes or alters a copyright notice with fraudulent intent, or makes a false representation of a material fact in a copyright registration application or other statement in connection with the registration.

In a civil suit brought by a copyright owner the damages awarded against the infringer go to the copyright owner. In a criminal proceeding brought by the U.S. Government, fines levied against an infringer are paid to the U.S. Government; the copyright owner does not share in them.

An "international copyright," which will provide the owner with protection in both the United States as well as in all foreign countries, does *not exist*. Like patents and trademarks, copyright protection is governed by the national laws of each country. Most countries do provide protection for foreign works if certain conditions are met. If you create or are responsible for the creation of a work for which copyright protection is available in the United States and you feel that it is essential or desirable to obtain foreign copyright protection, e.g., by virtue of your desire to distribute the copyrighted work abroad, you should immediately consult a copyright lawyer to determine the steps to be taken to obtain copyright protection in the country or countries of interest. At the very least, you should provide on the title page of all copies of the copyrighted work the form of copyright notice used in the United States, being sure, however, to use the symbol © and not "Copr." or "Copyright." In addition, you should place below the copyright notice the legend: "All rights reserved, including the right to reproduce this book or portions thereof in any form." By virtue of certain international agreements entered into by the United States and a number of foreign countries, placement of the foreign form of copyright notice and the above legend on the title page will provide rights under the copyright laws of the foreign countries who are parties to the agreements, as well as provide rights under the laws of the United States.

The Semiconductor Chip Protection Act of 1984

Prior to the adoption of this Act, which adds sections to the 1978 Copyright Act, protection of the circuits of integrated circuit chips was virtually nonexistent. Copyists were virtually free to produce an exact copy by analyzing the original. Due to the importance of this industry, i.e., about 14 billion dollars of domestic sales in 1984, and to provide meaningful protection, the Copyright Act was amended to provide for copyright registration of the "mask work," i.e., the mask from which the chips are made. A broad definition of mask work is provided so that a mask per se is not required. The use of laser etching to provide the copy would not avoid infringement of the copyrighted mask work.

There are several unique aspects in the Chip Act regarding infringement. Two of the most significant are that innocent purchasers of infringing chips are not liable until they have notice and for any use thereafter they are only liable for a reasonable royalty. Secondly, there is no liability where the mask work is created by reverse engineering. As to what is permissible, reverse infringing will undoubtedly be the subject of substantial litigation.

20

Trademarks

As noted in Chapter 1, intellectual property can be divided into four principal categories: (1) patents, (2) trademarks, (3) copyrights, and (4) trade secrets. Although the various kinds of intellectual property have some common attributes and in some respects can be considered to overlap one another in different respects, the basic nature of each is fundamentally quite different. In this chapter you will learn about trademarks.

The Trademark

Perhaps the logical starting point would be to define what a trademark is, namely, a word, name, symbol, device, or combination thereof which is adopted and used by a manufacturer, merchant, or organization on or in connection with his goods or merchandise to uniquely identify the source of his goods or merchandise and distinguish them from those of others. As you can see, then, the primary function of a trademark is to indicate the source or origin of the goods or merchandise.

For example, when you see a television with the mark Sony conspicuously located on it, you know it is marketed by a particular company, namely,

Sony Corporation, rather than by one of the many other companies that market televisions such as General Electric Company, Zenith Corporation, etc. Similarly, if you go to a drug store to purchase a headache remedy and you see aspirin in a jar with the trademark Bayer on it, you know that this particular brand of aspirin is marketed by a specific company, in this instance, Sterling Drug Company. As you may have already surmised, the trademark a company uses on its product may be the same as the company's corporate name, which is technically known as a *trade name*, as in the case of the Sony television which is marketed by Sony Corporation. However, this need not be the case and often is not, as is demonstrated by the fact that Bayer aspirin is marketed by the Sterling Drug Company. In some cases, a company's trademark is a varient, e.g., an abbreviation, of the company's trade name or corporate name. Illustrative of this is the mark SOHIO for Standard Oil of Ohio.

A company may use the same trademark on a variety of different products it manufactures. For example, General Electric Company utilizes the trademark General Electric on a variety of products including toasters, televisions, electric motors, etc. By the same token, a company may sell products which are quite similar, and in some cases virtually identical, under different trademarks. For example, Procter and Gamble Company sells a variety of different cleansing products, each with its own separate trademark, such as Ivory, Dawn, Comet, Tide, Cheer, etc.

A company may also use multiple trademarks on its products. For example, IBM Corporation has an IBM 360 computer, an IBM Selectric typewriter, an IBM Ramac random access memory, and an IBM Executary dictating machine. In each case, the product conspicuously bears the common trademark IBM and the specific trademark, e.g., Ramac, 360, Executary, Selectric, etc., which IBM Corporation reserves for and is unique to the particular product line.

From the foregoing examples, one thing is clear about trademarks. When you see a product with a particular trademark on it, you know it is marketed by a particular company. Usually, you know the identity of the company, although this is *not* a necessary incident of a trademark, and if the product or package bore no other marking on it except the trademark, there would be nothing defective about the trademark or its use.

In addition, the fact that a company markets a product under its own trademark does not necessarily mean that the company actually manufactured it. It is common knowledge that certain tools sold by Sears, Roebuck and Co. bearing the Craftsman trademark are not manufactured by Sears but are manufactured to Sears' specifications by one or more different companies which are completely separate and independent of Sears. In each case, however, the only trademark on the product is the Craftsman trademark; the products do not bear the trademark of the actual manufacturer. Incidentally, the company which manufactures tools which are sold by Sears under the trademark Craftsman may also manufacture and sell the same product under its own trademark. Thus it is conceivable that you could buy the exact same product manufactured by the exact same company from two different companies under two different trademarks and quite possibly never know it.

In addition to words, a trademark may consist of a symbol, such as the Mercedes-Benz three pointed star for automobiles used on its hubcaps and as a hood ornament, and the red cross symbol for services of the American National Red Cross organization.

Service, Certification, and Collective Marks

Closely related to trademarks, but in some respects different, are service marks, certification marks, and collective marks. Trademarks are applied to goods. Service marks are used to distinguish one's services from those of others. Collective marks and certification marks are used by merchandisers and/or organizations to certify a product as having met a certain standard or as coming from a certain geographic area, as the case may be.

Keeping in mind the function of a trademark in connection with products manufactured or sold by a company, you can readily understand the nature of a service mark if you just realize that a service mark is the same as a trademark, except that it is used in connection with the sale or rendering of services rather than products. Typical examples of service marks are Kelly Girl and Manpower for temporary help, Hertz and Avis for car rental, Holiday Inn and Ramada Inn for motel and dining services, Eastern Airlines and TWA for air transportation, and General Electric and Westinghouse for repairing and rewinding large electric motors.

Incidentally, a company may use the exact same mark both as a trademark and a service mark. For example, General Electric and Westinghouse are used as service marks for motor repair services and as trademarks for toasters.

Certification marks are used on products to indicate that they meet certain standards or come from a certain geographic location. For example, UL and the Seal of Approval of *Good Housekeeping* magazine, which appear on diverse products of many different companies, indicate that the products bearing the marks UL or the Good Housekeeping Seal of Approval have met the standards set by Underwriter's Laboratory or *Good Housekeeping* magazine, which own the respective marks. Similarly, the mark Roquefort on cheese certifies that the cheese is from the Roquefort district of France. Fabrics bearing the mark Sanforized indicate that the fabrics were subjected to a certain antishrink process designed to minimize shrinkage when laundered; similarly, dry cleaners using the mark Sanitone use a particular dry-cleaning process.

Finally, collective marks, which can also appear on diverse products of different manufacturers, indicate membership or sponsorship by a particular organization. For example, articles of clothing manufactured by companies having unionized employees belonging to the International Ladies Garment Workers Union often bear the mark ILGWU to indicate that the products were manufactured by persons belonging to the union. This is the collective mark of the International Ladies Garment Workers Union. Collective marks can also denote membership in an organization, such as AFL-CIO for union locals which are member locals of the parent union, American Federation of Labor–Congress of Industrial Organization, or Tau Beta Pi, which is an engineering honor society.

Securing Rights in Marks

A person, organization or company, or group obtains rights in a mark—whether it be a trademark, service mark, certification mark, or collective mark—only by actually (1) using the mark on or in connection with goods or services which (2) move in commerce. The commerce requirement is satisfied regardless of whether the goods or services are involved in interstate, intrastate, or foreign commerce involving the United States and a foreign country. As for the use requirement, if the mark is a trademark, the mark generally must be affixed to the goods or to their packaging. In certain limited situations where it is impractical to actually affix the mark to the goods or to the packages, such as when pickles of a certain brand are sold individually from a large pickle barrel by a delicatessen, it is sufficient if the mark appears on the pickle barrel which contains the pickles. Mere use of the mark in printed advertising, radio or TV ads, or like promotional material is not, in and of itself, sufficient if the mark is not actually used on the goods themselves, their packaging, or in close association with the product such as on the barrel from which pickles are sold. In the case of service marks, where there is no product per se but rather only a service, it is sufficient if the mark is used in advertising, promotion, or on the letterhead or door sign of the company rendering the services. Similarly, rights in certification marks and collective marks arise when they are actually used in whatever manner is appropriate to the particular type of mark.

U.S. Trademark Registrations

As you have seen, trademark rights can only vest in the owner after use of the mark in commerce on or in connection with the goods or services being offered by the owner. However, once the owner has used the mark and thereby acquired trademark rights to it, nothing additional need be done to vest those trademark rights. However, when a mark has been used on goods or services in interstate or foreign commerce, vis-a-vis only in intrastate commerce, there is a procedure available by which the owner of a trademark can register the fact that he does own a particular mark, the registration being made at the U.S. Patent and Trademark Office in Washington, D.C. A typical trademark registration is reproduced at Appendix E.

Although it is not necessary for an owner to register his claim to a particular trademark used in interstate or foreign commerce with the U.S. Patent and Trademark Office, there are a number of advantages in doing so. Principal among these:

1. Following five years of continuous use of the mark after registration, the federal registrant's right to continue to use the mark becomes incontestable. Incontestability means that the certificate of registration is conclusive evidence of ownership and the right to use of the mark. An incontestable registration may still be contested but only on a limited number of grounds. It may not be contested on the basis that the mark is descriptive.
2. The owner of a federally registered trademark can stop the importation of goods bearing his mark by others who have not been given permission to use the mark.

3. Federal registration puts everyone in the United States on notice of your claim to ownership of the mark. A subsequent adopter of the mark for the same or similar goods, even though not actually aware of your registration, is precluded from obtaining valid trademark rights in your marketing area or in any natural zone of expansion of it. If you adopt the mark in one section of the contry but do not register it with the U.S. Patent and Trademark Office and a person in another part of the country innocently and without knowledge of your mark subsequently adopts the same unregistered mark for the same or similar goods, then you would not be able to stop him from using the mark in his marketing area. Thus, should you later decide to expand marketing of your nonfederally registered trademarked product, e.g., from the New York metropolitan area to the Boston metropolitan area, a subsequent innocent adopter of the mark in Boston could not be made to stop using the mark when you enter his market area. Moreover, he could preclude you from marketing your product in his area using the trademark which he innocently adopted after you selected the mark.
4. Certain additional advantages of a federally registered trademark exist, primarily of a procedural nature, such as the right to bring suit against a trademark infringer in a U.S. federal district court instead of in a state court, the availability of certain remedies against an infringer, and the like.
5. A federal registration also provides additional legal remedies and relief in the event that counterfeit products bearing the registered trademark are being sold.

Registration procedures also exist in the individual states. State registration provides some limited advantages, on a statewide basis, over non-registration.

Selection of a Mark

If you have the opportunity to select a mark for a product or service, the more fanciful and arbitrary the mark, the stronger the protection which usually results. For example, coined words such as Kodak, Xerox, etc., which have absolutely no dictionary meaning, are very distinctive and afford the owner a relatively wide range of protection against would-be infringers who would attempt to market similar products using the Kodak or Xerox mark or a variant thereof. On the other end of the distinctiveness spectrum are what are known as descriptive marks. A descriptive mark is a conventional dictionary word, the meaning of which describes a function, quality, or characteristic of the product, such as "Nutritious" for fertilizer. Although it is possible to obtain valid trademark rights to a descriptive trademark, it is much more difficult to do so. Before trademark rights will vest in the user of a descriptive mark, it is necessary to show that the mark has acquired what is known as *secondary meaning*, i.e., the purchasing public has come to identify the particular mark, such as Nutritious, as the mark of a particular fertilizer manufacturer rather than merely as a descriptive term or synonym for fertilizer. Even where secondary meaning has occurred, others may be able to use the word *nutritious* to describe a fertilizer in a descriptive sense but not use it in a trademark sense. For example, competitors of the fertilizer manufacturer who has acquired valid trademark rights for the mark Nutritious for fertilizer could not stop others from including in their promotional literature statements that their fertilizers "contain nutritious elements." He could, however, stop them from calling their products "nutritious" fertilizer.

Between the extremes of the arbitrary or fanciful mark, such as Kodak, and the descriptive mark, such as Nutritious, are what are known as suggestive marks. A suggestive mark is one which does not really describe a characteristic of the product but rather only suggests what the characteristic is. For example, Seventeen suggests, but does not describe, a fashion magazine for teenage girls. Suggestive marks are the most popular since they are "catchy" and therefore more easily remembered. In addition, they do suggest to a potential purchaser first hearing the mark what the product is. This is in contrast to the owner of an arbitrary mark, such as Kodak, who must first educate the public with regard to the products marketed under the mark.

Not all marks which a company could adopt give rise to valid trademark rights or are registrable. For example, marks which are immoral, deceptive or scandalous, or which disparage or falsely suggest a connection with persons (living or dead), institutions, beliefs or national symbols, or which will bring them into contempt or disrepute cannot be registered. Marks which consist of or comprise the flag or coat of arms or other insignia of the United States or any state or municipality or any foreign nation or any simulation thereof also are refused registration. Similarly, marks which consist of or comprise a name, portrait, or signature identifying a particular living individual (except by his written consent), or the name, signature, or portrait of a deceased president of the United States during the life of his widow (except by written consent of the widow) are not entitled to registration. Of course, no mark can be registered if it so resembles a subsisting mark used by others that when it is applied by you to your goods it will cause confusion, mistake, or deception.

The trademark laws of the United States are different from most foreign countries with respect to the manner in which a person acquires a trademark. In the United States, as indicated, a person acquires the rights to a mark by actually using it on or in connection with his goods or services. Until this has been done, the person has acquired no trademark rights and is unable to obtain a federal or state trademark registration. By way of contrast, in most foreign countries a person can acquire trademark rights merely by filing an application for the trademark with the appropriate governmental agency. Thus, in a number of foreign countries a resident of such country may obtain rights to a trademark, having never used it. When you attempt to import your goods into his country bearing the mark he has registered but never used, he can prevent it. Since the laws of the various foreign countries are all different, if you are interested in selecting a mark for a product to be exported to a foreign country, you should first have a search made by your trademark lawyer to determine whether the mark is available in that country.

Finally, your trademark lawyer's aid in investigating the availability of a proposed mark should be solicited before the mark is finally adopted. This could avoid the making of large expenditures in generating goodwill in a mark, only to later find out another party acquired dominant prior rights in the same mark. See the final section of this chapter.

Using the Trademark

To avoid losing rights to a trademark which you have once acquired by adopting and using the mark on or in connection with your goods or services, certain precautions must be taken to assure that the mark continues to distinguish your goods from those of others:

1. Use the mark as a proper adjective modifying the descriptive or common name of the product, such as, Xerox copier, IBM typewriter, Manpower temporary help, Vaseline petroleum jelly. Do not use the mark as a synonym for the product, e.g., "Vaseline is a great balm" or "Put a Band-aid on your cut."
2. Do not bastardize the mark by using prefixes or suffixes with it, abbreviating it, or using it as a verb or in plural form. For example, it is improper usage to "Xerox the page"; one does, however, "make a Xerox copy of the page" or "make a copy on the Xerox copier." It is also incorrect to describe one who is using a Xerox copier as engaged in "Xeroxing." There are "two IBM Executive typewriters," not "two Executives."
3. When using the mark in print, use distinctive type, italics, quotation marks, or colors different from the rest of the type to set the mark off from the remainder of the text.
4. Use a trademark notice with the mark to advise others it is a mark. For example, follow the mark with the symbol ® as a superscript if the mark is federally registered or a superscript TM or SM if the mark is not a federally registered trademark or service mark. Alternatively, note in a legend on the page on which the mark appears that it is a trademark of XYZ Company for its products.

If the foregoing precautions are not taken, you run the risk that the mark will become what is known as *generic*, i.e., your mark will become the common descriptive term for the general type of goods you market rather than function to distinguish your specific product from those of other suppliers. Cellophane and aspirin, once valid trademarks of duPont and Bayer for their specific brand of transparent wrap and headache remedy, became such common substitutes for the descriptive terms "transparent wrap" and "headache remedy" that the marks no longer functioned to distinguish the specific products of duPont and Bayer from those of others. When this occurs, the marks become generic and trademark rights cease.

Trademark Infringement

One infringes another's trademark by unauthorized use of the mark in such a way that it creates confusion in the mind of the public with respect to the source of the goods or services of the unauthorized user, e.g., causes the public to believe that they are those of the trademark owner when in fact they are not. This is what is known as "palming off." In effect, the unauthorized user is "riding on the commercial coattails" of the trademark owner.

Since a trademark is used by a company to distinguish its own products from those of its competitors, it is improper—and in fact constitutes unfair competition—for one company to utilize on its product a mark which causes confusion in the minds of the public with respect to the origin of the goods. Confusion as to the source of goods can arise by reason of the similarity of the marks themselves or similarity of the goods. If the goods are completely different and move through entirely different marketing channels, such as automobiles and dog food, then each manufacturer could conceivably use the exact same trademark, such as Cadillac, and there would be no confusion. Consumers seeing Cadillac dog biscuits on the shelf of a supermarket would not likely be confused and think they originated from the same company that sells Cadillac automobiles.

However, as the nature of goods converges, i.e., becomes more similar, in order to avoid confusion as to origin which gives rise to trademark infringement, the marks must diverge, i.e., become more dissimilar. A good example of trademarks for the same type of goods which give rise to no confusion as to origin are Kodak and Ansco for photographic film. On the other hand, the trademarks Country Kitchen and Country Pantry, when applied to salad dressings of different companies, are likely to create confusion in the minds of purchasers. Both trademarks utilize the same word, *country*, as the first word of the trademarks, and the second words *kitchen* and *pantry*, although different, have the same connotation.

The degree of similarity between trademarks which can exist before confusion in the minds of the consuming public sets in depends not only on the similarity or dissimilarity of the products and the marks themselves, but also on certain other factors, such as the discrimination likely to be exercised by the buyer when selecting one brand over another. This is in part dependent on such factors as his education, age, and sophistication. Also a factor is the cost of the product. Commonplace items, such as candy bars, different brands of which are marketed through the same channels such as retail supermarkets, are low-cost items. They are purchased on impulse by people who are not necessarily particularly discriminating. In such situations, greater divergence in the marks themselves is required to avoid confusion as to the product origin than is needed, e.g., between marks used on $2 million computers for which the purchase decision is made by a highly educated, trained, and sophisticated business executive, and then only after very careful consideration.

Sometimes, even though the products are quite similar and the marks are quite similar, there may be no confusion if the channels of distribution for the trademarked products are completely different. For example, two different companies may market the same type of detergent under quite similar trademarks without any confusion arising as to origin of the respective products. For example, one of the companies may market its product to physicians and hospitals for cleansing parts of the human body prior to performing surgical operations, whereas the other company markets its product to industrial users for removing grease from metal extrusions. Such divergence in marketing channels may obviate any possibility of confusion.

Trademark infringement can result even when the products or services of the trademark owner and unauthorized use of the mark convey a false impression to the public that the unauthorized user is associated with or sponsored by the trademark owner. For example, if a manufacturer of clothing for teenage girls elected to use the mark Miss Seventeen without authority of the publisher which owns the mark Seventeen for a fashion magazine directed to teenage girls, trademark infringement would result because the public would erroneously think the clothes marketed under Miss Seventeen are associated with or sponsored by the publisher of Seventeen magazine.

The most flagrant type of trademark infringement is counterfeiting. This exists when the infringer sells a product substantially identical in appearance to the legitimate product and with a trademark substantially identical to the legitimate trademark. In most instances the counterfeit is inferior to the legitimate product. Product trademarks ranging from ones for famous-brand motor oil to ones for expensive designer-named garments have been counterfeited. In response to the increasing sale of counterfeit merchandise, Congress amended the Federal Trademark Act to provide that in the case of a federally registered trademark, a court may impose increased damages, criminal penalties and issue seizure orders for the counterfeit goods without prior notice.

The prohibitions against counterfeit goods do not extend to "gray market" goods. In the typical gray market situation, there exists a gray market product that is imported into the United States with a trademark and a similar product, i.e., a domestic product that bears the same trademark. The quality of the two are the same and the products may be indistinguishable. The owner of the U.S. right to the use of U.S. trademarks is different from the owner of the foreign trademark being used on the imported product, although there may be a parent-subsidiary relationship. Gray market goods are usually bought abroad at prices which permit them to be resold in the United States at a price below the sale price of the domestic product. Obviously, the foreign exchange rates have a significant effect on this market.

The owner of the U.S. trademark rights *may* be able to block the importation of the gray market product. The case law in this area is still developing and a number of factors bear on the question.

Trademark rights may be litigated in federal or state courts, depending on whether the mark is federally registered and/or the parties are citizens of different states. Moreover, if importation is involved, a proceeding may be instituted before the International Trade Commission which can lead to an award of an injunction to block the importation. The right to block importation may also be obtained from the U.S. Customs.

Trademark Investigations

The consequences of trademark infringement can be serious. An infringer can be held responsible for damages and attorneys' fees and/or subject to an injunction prohibiting further use of the mark. Because of these factors, since marks can be so valuable, and since development of good will in marks can be expensive, it is normally advisable to have a trademark lawyer investigate any proposed mark prior to its adoption. Such an investigation will be directed to determining whether the mark is free for use as proposed without infringing the trademark rights of others. It should also be determined whether the mark can be registered or at least whether it can function as a distinctive, protectable mark. Having this information, a company can make a more informed decision as to whether a proposed mark should be adopted.

There are two types of trademark searches that are normally performed before adoption and use of the mark. The first is limited to federal registrations and applications. This type of search only provides a limited amount of information as to the questions of whether one may use and register the mark. This is because one who is using the same or legally similar goods may block the use and registration even though a federal registration has not been obtained. This is why there is a second type of search which is broader in scope and which seeks to locate users of conflicting unregistered trademarks. Telephone directories, trade directories, company name listings, and state trademark registrations are the principal areas searched. Several companies maintain computer data bases with such information and, for a fee, will search a trademark.

Trademark searches performed before adoption and use are extremely important. It is *not* sufficient to merely incorporate under a name which will be used as the trademark. This does not give one the right to use the name as a trademark.

There has been a tremendous expansion in the field of trademark licensing since the 1960s. Franchising, which includes, among other things, a trademark license to the franchisee, has increased by billions of dollars. The trademark is the cornerstone of any franchise. Careful selection, proper adoption and use, and control over the use of the mark are very important. The mark adopted should be one which can be federally registered and should be one which can be used in all geographic areas without infringing any other's trademark rights. To protect the validity of the trademark, the licensed use must be controlled by the licensor. Normally this means that the licensor must control the nature and quality of the goods or services on which the mark may be used. If the use is not controlled, the trademark registration can be cancelled.

In addition to franchising, the licensing of trademarks has mushroomed in the area of "collateral product" trademark licensing. This normally takes the form of a trademark owner, usually of a famous trademark, licensing the use of the mark on other products. For example, the owner of, say, the famous XYZ beverage trademark may license another to sell articles of clothing under the XYZ trademark. Again, as with franchising, the nature and quality of the collateral goods on which the licensed mark is used must be controlled.

Appendixes

Appendix A

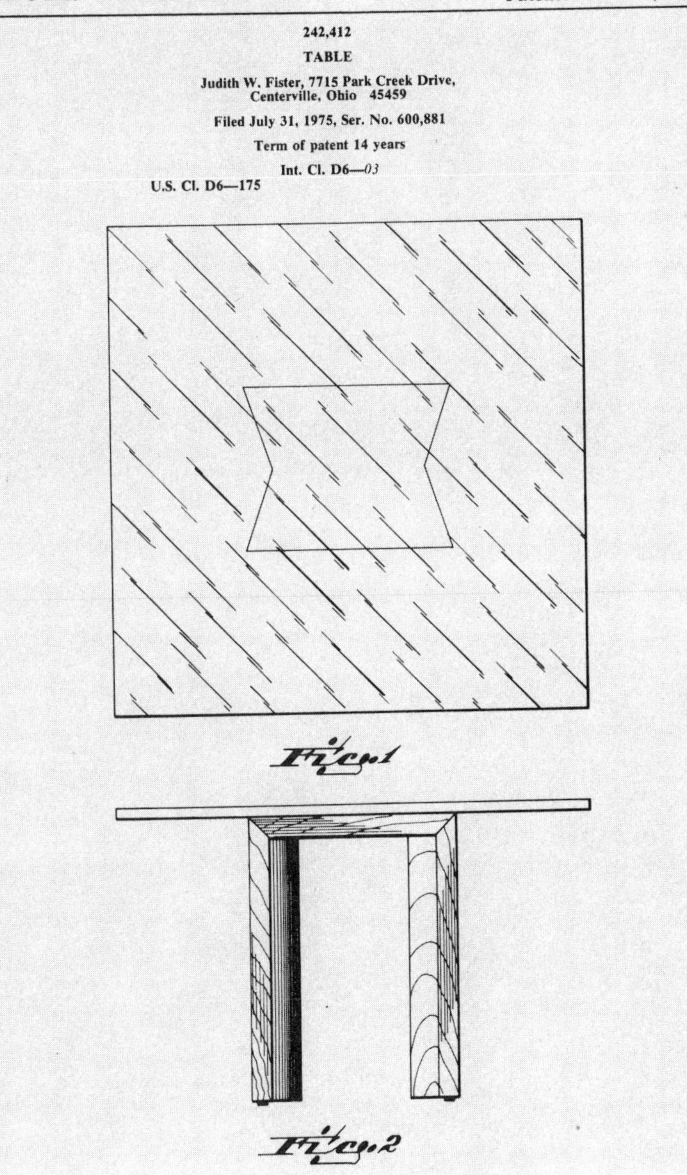

United States Patent

Des. 242,412
Patented Nov. 23, 1976

242,412
TABLE

Judith W. Fister, 7715 Park Creek Drive,
Centerville, Ohio 45459

Filed July 31, 1975, Ser. No. 600,881

Term of patent 14 years

Int. Cl. D6—03

U.S. Cl. D6—175

Fig. 1

Fig. 2

Des. 242,412
Page 2 of 2

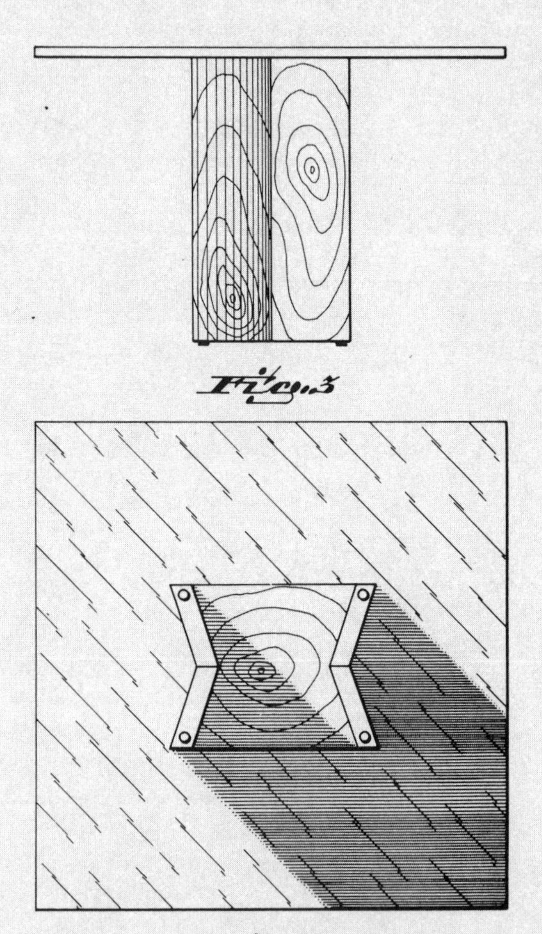

FIG. 1 is a top plan view of a table showing my new design;

FIG. 2 is a side elevational view thereof, the other side being a mirror image;

FIG. 3 is an end elevational view thereof, the other end being a mirror image; and

FIG. 4 is a bottom plan view.

I claim:

The ornamental design for a table, as shown and described.

References Cited

UNITED STATES PATENTS

D. 222,278	10/1971	Thornhill	D6—177
D. 225,950	1/1973	Kienel	D6—177
D. 236,926	9/1975	Alexander	D6—177
D. 238,407	1/1976	Owens	D6—177 X

BRUCE W. DUNKINS, Primary Examiner

Appendix B

United States Patent [19]

Greenwell

[11] **3,996,723**

[45] **Dec. 14, 1976**

[54] **ARTICLE COLLATOR**

[75] Inventor: **Joseph Daniel Greenwell,** Florence, Ky.

[73] Assignee: **R. A. Jones & Company, Inc.,** Covington, Ky.

[22] Filed: **Mar. 28, 1975**

[21] Appl. No.: **563,095**

[52] **U.S. Cl.** .... **53/59 R;** 53/62; 53/164; 214/6 DK

[51] **Int. Cl.²** **B65B 57/10;** B65B 57/20

[58] **Field of Search** .... 53/59 R, 62, 164; 198/21; 214/6 DK

[56] **References Cited**

UNITED STATES PATENTS

3,193,078	7/1965	Amenta et al.	53/59 RX
3,512,336	5/1970	Rosecrans	53/164
3,641,735	2/1972	Daily et al.	53/164
3,729,895	5/1973	Kramer et al.	53/59 R
3,817,368	6/1974	Wentz et al.	198/21

Primary Examiner—Travis S. McGehee

Attorney, Agent, or Firm—Wood, Herron & Evans

[57] **ABSTRACT**

A collator having an upper receiver for receiving a layer of articles, a lower receiver for accumulating multiple layers of articles and trap doors for the receivers. A swinging conveyor is pivoted at its upper end and means are provided to swing its lower end to any of a plurality of discharge points adjacent said upper receiver. Control means including counters are provided to count articles as they pass from the swinging conveyor to the upper receiver and when a preselected number is received by the upper conveyor, its trap door is operated to drop that group into the lower receiver. Several such drops are made until the desired total is accumulated, whereupon the trap door for the lower receiver is opened to drop the accumulated group into a receptacle.

12 Claims, 15 Drawing Figures

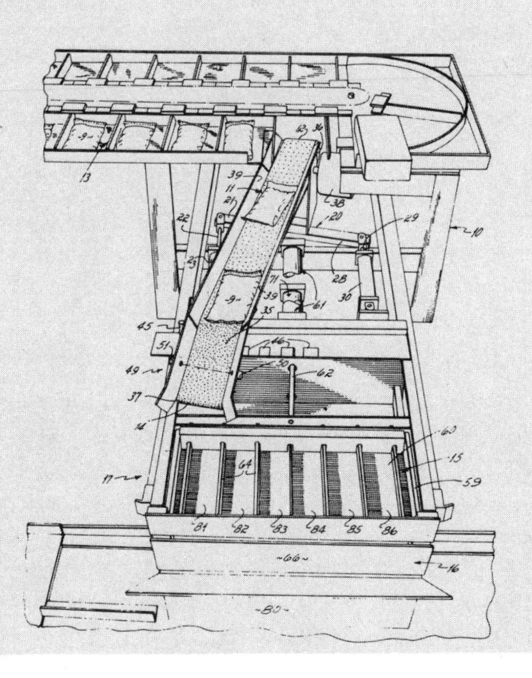

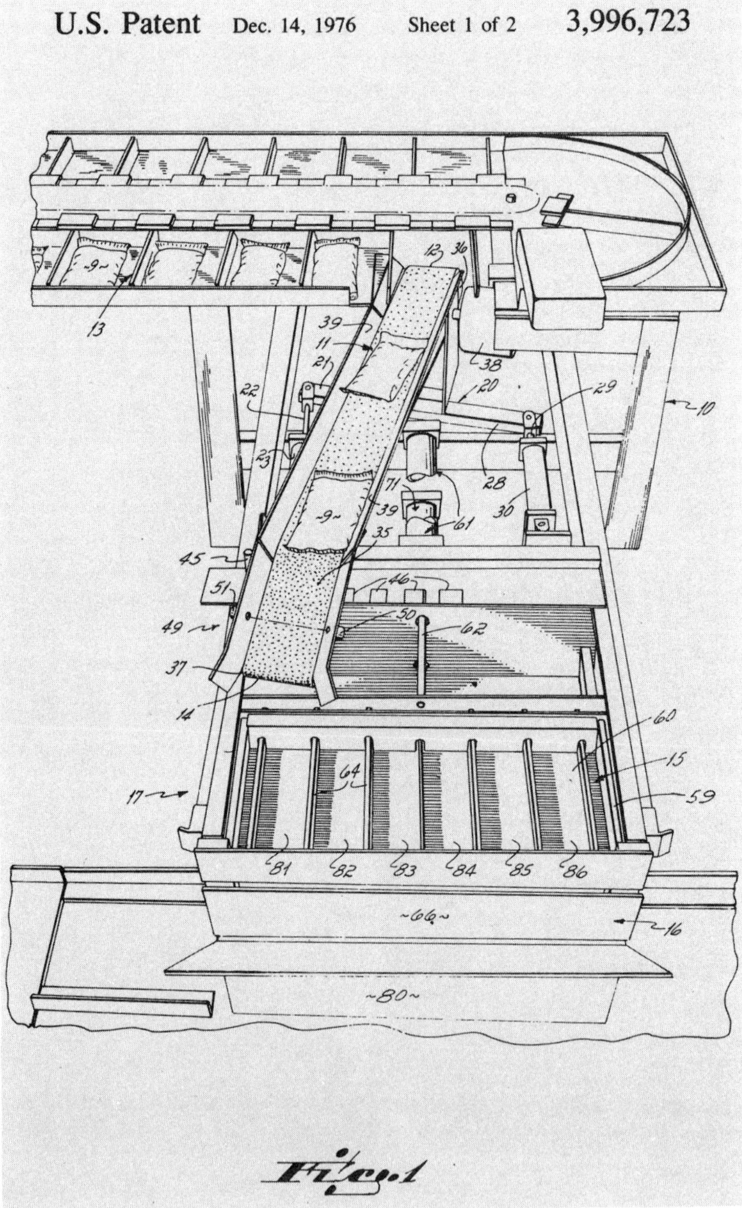
U.S. Patent Dec. 14, 1976 Sheet 1 of 2 3,996,723
9
13
12
36
39
11
21
22
23
38
20
29
10
28
71
61
30
9
39
35
45
51
46
50
62
49
37
14
60
15
64
59
17
81
82
83
84
85
86
66
16
80

Fig. 1

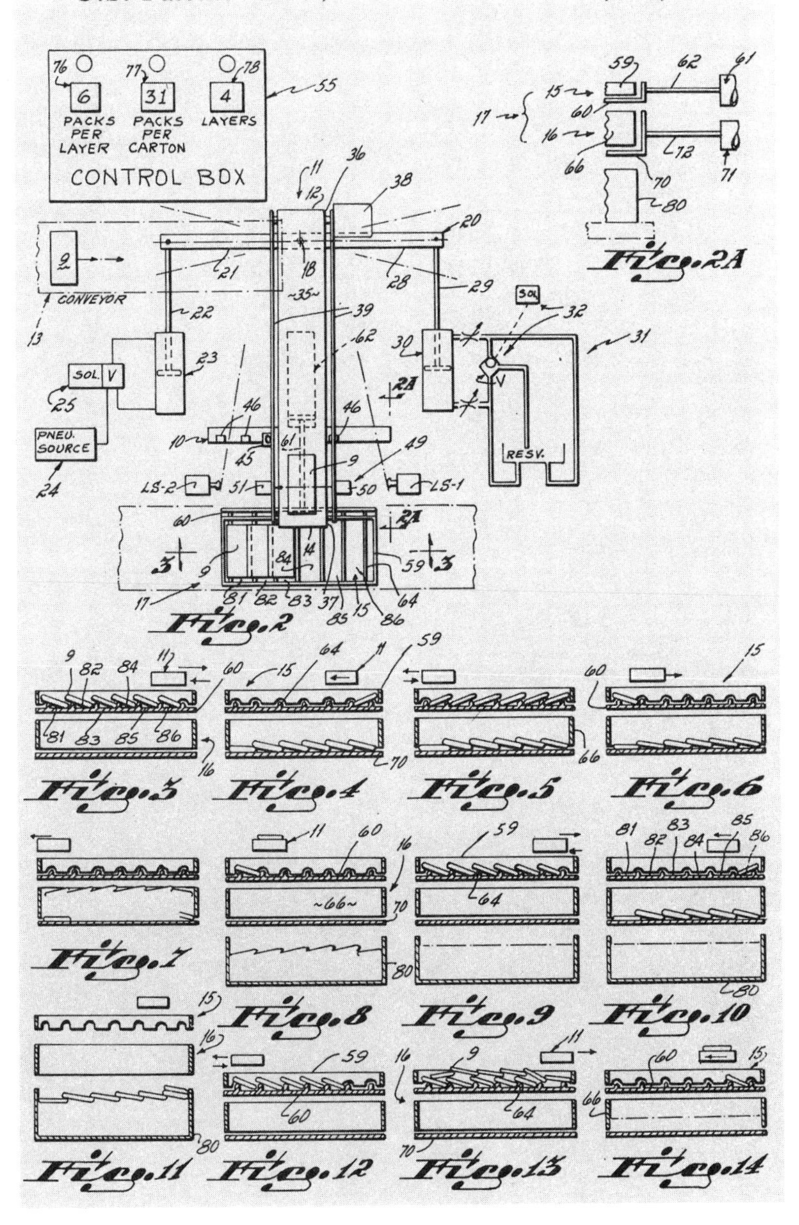
U.S. Patent Dec. 14, 1976 Sheet 2 of 2 3,996,723
6
PACKS PER LAYER
31
PACKS PER CARTON
4
LAYERS
CONTROL BOX
CONVEYOR
SOL.
V
PNEU. SOURCE
RESV.
LS-2
LS-1
Fig. 2
Fig. 2A
Fig. 3
Fig. 4
Fig. 5
Fig. 6
Fig. 7
Fig. 8
Fig. 9
Fig. 10
Fig. 11
Fig. 12
Fig. 13
Fig. 14

3,996,723

1

ARTICLE COLLATOR

This invention relates to a collator by which articles, individually conveyed to the collator, are assembled in layers and deposited into a receptacle.

In commerce there are many instances when it is desired to pack articles in a more or less organized fashion into cartons. In packing them, it is necessary to count them so that each carton has the same number and to organize them into layers so that they can be efficiently and economically packed in cartons more or less precisely sized for the number of articles to be packed.

These counting, collating and packing operations are normally performed manually simply because of the fact that there has not been commercially available equipment which can satisfactorily perform all of the operations and still have the degree of versatility to accommodate varying numbers of articles as well as various sizes and shapes. A few patents have disclosed collators, but it appears that none has achieved any commerical recognition.

An objective of the present invention has been to provide a collator adapted to receive articles one at a time from a conveyor and to organize the articles into a group of several layers of the articles of the desired number and to deposit the articles into a receptacle such as a carton or a product bucket.

Another objective of the invention has been to provide a collator which can count, collate and pack articles rapidly

Another objective of the invention has been to provide a collator which is versatile in that it can accommodate varying numbers of articles, varying layers of articles, as well as various sizes and shapes of articles.

To achieve these objectives, the invention provides a collator having a swinging conveyor which receives articles one at a time. Means are provided to count the articles passing along the conveyor and to swing the discharge end of the conveyor to the next succeeding discharge point as each article is discharged. Underlying the discharge end of the conveyor are upper and lower receivers each having trap doors. The upper receiver has a plurality of compartments into which the conveyor discharges articles one at a time. When a first group of a preselected number of articles is received by the upper receiver, that group is dropped into the lower receiver. That operation is continued until the desired total number is accumulated, whereupon the lower trap door is opened to drop all articles into a receptacle such as a product bucket or a conventional cartoner or a tray-type carton.

A detector, preferably a photocell, is mounted on the discharge end of the conveyor to form a part of the counting means. It provides a "demand" type operation, that is, the various functions of the apparatus are performed only as determined by the actual flow of articles from the conveyor into the receivers.

The apparatus includes settable counters for the total number of packs, the number of packs in each layer, and the number of layers. This feature imparts considerable versatility to the collator in enabling it to accommodate a wide variety of numbers of articles to be packed and a variety of sizes of articles to be packed and shapes of cartons into which they are packed.

The trap door of the upper receiver is provided with a plurality of spaced parallel ribs to define compartments. The ribs also serve to cause articles which are wider that the compartment widths to assume a shingled orientation, thus adding to the versatility of the collator to accommodate articles of various sizes.

2

While the double drop apparatus using two receivers is not per se novel, its use in the present collator combination is particularly desirable where such high speeds are desired that it might not be possible to drop a complete accumulation of articles before the first article of the next group appears at the discharge end of the conveyor.

The present invention is suitable for intermittently operated cartoners or for continuous motion cartoners, the latter requiring a longitudinally movable second receiver, that is to say, a second receiver which moves in the direction of the receptacle into which it is dropping its articles as it is dropping its articles as shown in copending application Ser. No. 491,313, filed July 24, 1974.

The several objectives and features of the invention will become more readily apparent from the following detailed description taken in conjunction with the accompanying drawings in which:

FIG. 1 is a perspective view, partially broken away, of the collator apparatus;

FIG. 2 is a diagrammatic plan view of the collator apparatus;

FIG. 2A is a cross sectional view taken along lines 2A—2A of FIG. 2; and

FIGS. 3–14 are cross sectional views taken along lines 3—3 of FIG. 2 illustrating the sequence of operation.

Referring to FIGS. 1 and 2, the apparatus is mounted on a base **10** and includes a swinging conveyor **11** which receives articles at its upstream end **12** from a feeding conveyor **13**. The conveyor **11** has a downstream end **14** overlying an accumulator **17** which has an upper receiver **15** and a lower receiver **16**. The conveyor **11** is mounted to pivot about an axis **18** at its upstream end so that its downstream end swings to any one of a number of discharge points overlying the upper receiver **15**.

The conveyor **11** is mounted on a crossbar **20** which is adapted to pivot about the axis **18**. The crossbar has a first laterally projecting arm **21** which is connected to a piston rod **22** of a pneumatic double acting piston and cylinder **23**. The double acting piston and cylinder **23** is operated from a source of air **24** under pressure through a solenoid actuated valve **25**. The piston and cylinder **23** is the driver for the swinging conveyor **11**. The crossbar **20** has another arm **28** projecting laterally to the side opposite arm **21** and which is connected to a rod **29** of the piston in a hyrdaulic double acting piston and cylinder **30**. The double acting piston and cylinder **30** functions as a brake to stop the conveyor **11** at each of the desired discharge points. The double acting piston and cylinder **30** is connected to a hydraulic circuit **31** including a solenoid operated valve **32** which permits circulation of fluid when the conveyor is being driven by the piston and cylinder **23** and which blocks the circulation of fluid when the conveyor is to be braked at a discharge point. As an alternative to the two pistons and cylinders, the conveyor could be swung by a pulsed stepping motor.

The conveyor **11** has an endless belt **35** passing around sheaves **36** at the upstream end and **37** at the downstream end. The sheave at the upstream end is driven by a motor **38** mounted on the crossbar **20**. A pair of longitudinally extending guides **39** are mounted at the conveyor on each side of the belt **35** and project above the surface of the belt to confine the articles to the belt as they pass from the conveyor **13** onto the conveyor **11** and then into the upper receiver **15**.

An electric eye **45** is mounted on one of the guides **39** and is directed toward blocks of reflective indicia **46** mounted on the base **10**. Each of the indicia **46** corresponds to a discharge point at which the conveyor **11** should be braked. A beam of light reflecting off these indicia into the electric eye **45** determines the angular location of the conveyor **11** and signals the solenoid operated valve **32** to brake the conveyor **11**. The blocks of indicia are of substantial size and will trigger the brake at differing discharge points depending upon whether the conveyor is swinging left or right. This permits a useful orientation of the articles in the upper receiver compartments, as will appear below.

At the downstream end of the conveyor is an article detector **49** consisting of a light source **50** and an electric eye **51**. The electric eye **51** generates a signal each time an article **9** interrupts the beam from the source **50** as the article **9** passes over the discharge end of the conveyor **11**.

A limit switch LS-1 is on one side of the conveyor **11** and a limit switch LS-2 is on the other side of the conveyor **11**. The limit switches may be moved laterally with respect to the conveyor **11** to reduce the length of its excursion when, for example, packing conditions require less than the maximum number of articles in each layer. Thus, six articles per layer can be accommodated in the illustrated form of the invention. That number can be reduced as, for example, to five by shifting limit switch LS-1 toward limit switch LS-2 by a distance equal to the dimension of one compartment. When a limit switch is engaged by the conveyor, at the end of its arcuate excursion, it signals the conveyor swinging means, that is, the circuits for the pistons and cylinders **23** and **30** to reverse the direction of their operation.

The upper receiver **15** is formed partly by a fixed rectangular frame **59** which confines the articles as they are dropped into it. Below the frame **59** is a trap door **60** which is reciprocable by a pneumatic piston and cylinder **61** connected to it by a rod **62**. The trap door has a plurality of upwardly projecting spaced parallel ribs **64** which define compartments for the receipt of articles, six compartments being shown in the illustrated form of the invention.

The lower receiver **16** has a fixed rectangular frame **66** which is deeper than the rectangular frame **59** of the upper receiver, since the lower receiver will normally receive multiple layers of articles. Below the frame **66** the lower receiver has a trap door **70** which is reciprocable by a pneumatic double acting piston and cylinder **71** connected to it by a piston rod **72**.

The apparatus has a control located in the control box **55** which includes three settable counters **76**, **77** and **78**. The control circuit utilizes conventional logic to perform the simple counting and resetting operations described below. The counter **76** counts the packs per layer and when the preselected number of packs is counted effects the operation of the piston and cylinder **61** to open the trap door **60** of the upper receiver **15**. Each limit switch LS-1 and LS-2 is connected to the counter **76** to reset the counter to zero for reasons which will appear below. Counter **77** is settable to the total number of packs desired. When that number is counted, both pistons and cylinders **61** and **71** are operated to drop the total number of articles into a receptacle **80** underlying the lower receiver **16**. At this point the filled receptacle is moved out and a new receptacle brought into position and all counters are reset to zero to begin recounting.

Counter **78** counts the numbers of layers which have been dropped and its use is optional. When the preselected number of layers has been dropped, trap door **70** on the lower receiver is operated to drop the partial load into a receptacle. This counter does not have to be employed when the lower receiver is deep enough to accommodate a full load. Where the lower receiver is shallow, it might accumulate a substantial portion of the full load until filled. Thereafter, using the layer counter to drop the partial load, the operation can be continued since the lower receiver will then have space to accommodate additional layers.

The operation of the invention can be understood by reference to FIGS. 3–14 taken in conjunction with FIGS. 1 and 2. Let it be assumed that it is desired to have six articles per layer and **31** total articles in the carton. Further, let it be assumed that the lower receiver has only sufficient depth to accommodate four layers. Those determinations are entered into the control box with counter **76** being set at 6, counter **77** being set at 31 and counter **78** being set at 4.

The limit switches LS-1 and LS-2 are set the maximum distance apart since six articles in a layer is the maximum which the illustrated collator can accommodate.

At start-up, assume that the conveyor **11** overlies the first compartment on the left, as shown in FIG. 1. As the first article **9** passes the detector **49** and drops into the leftmost compartment indicated at **81**, the counters **76**, **77** record the fact that one article has passed into the receiver. The detector also signals the solenoid operated valve **25** to energize piston and cylinder **23** to swing the conveyor toward the next compartment **82**. As the conveyor **11** reaches the compartment **82**, electric eye **45** reads the appropriate indicia **46** and signals the solenoid operated valve **32** to brake the conveyor **11**. Because of the width of the indicia, the conveyor will be stopped close to the leading or left (as viewed in the drawings) rib **64** defining a compartment.

It will be noted that the article **9** is wider than the space between the ribs **64** on the trap door **60** which forms the bottom wall of the upper receiver **15**. Therefore, one edge of the article falls on the trap door **60** and the other edge of the article is slightly elevated by engagement of the article with the rib **64** so as to begin a shingled orientation of the articles.

The operation continues with the discharge of articles into compartments **82–86** until a layer of six has been deposited in the upper receiver **15**. This number of articles having been counted by the counter **76**, the trap door **60** of the upper receiver is pulled, by the piston and cylinder **61**, out of the receiver **15** to drop the layer of six articles into the lower receiver **16**. That first drop is counted by the layer counter **78**. The conveyor **11** continues to swing until it engages limit switch LS-1 which, when contracted, effects a reversal of the circuits of solenoid operated valves **25** and **32** so as to condition the conveyor **11** for swinging movement toward the left as viewed in FIG. 1.

The conveyor **11** then swings to a position overlying compartment **86** where it is braked and the first article of the second layer is deposited. The conveyor continues to swing toward the left and serially deposits articles in the compartments **86–81.** Note that the ribs **64** of the trap door **60** cause a shingling in the opposite direction as caused by the width of the indicia **46.** Thus, each article, which is wider than the compartment, will not be obstructed by the preceding article dropped into the preceding compartment.

When the sixth article is received, trap door **60** is again withdrawn to drop that group into the lower receiver, as shown in FIG. **6.** That excursion back and forth is continued until four layers have been dropped, as shown in FIG. **7.** Upon the dropping of the fourth layer, the layer counter **78** triggers the operation of the trap door **70** in the lower receiver to cause all accumulated (24) articles to drop into the receptacle **80** (FIG. **8**).

The operation continues to deposit a group of six more articles in the upper receiver (FIG. **9**) which is thereupon dropped into the lower receiver. On the return leftward excursion of the conveyor a final article (FIG. **10**) is dropped into upper receiver **15,** now making the total 31. When the total of 31 is counted by the counter **77,** both trap doors **60** and **70** are withdrawn to drop the remaining seven articles into the receptacle **80** (FIG. **11**).

At this point, the conveyor **11** has begun a leftward excursion and now overlies compartment **85.** It continues to deposit articles in a leftward excursion, dropping five articles into compartments **85** to **81** (FIG. **12**). When the fifth article is dropped into compartment **81,** the conveyor **11** continues to swing leftward until limit switch LS-**2** is engaged which resets the packs per layer counter **76** to zero. The conveyor will then swing back to overlie compartment **81** and continue to drop six more articles in the respective compartments **81–86** on top of the first five which were dropped. The accumulation of 11 articles in the upper receiver is permitted by virtue of the resetting of the packs per layer counter to zero after five were dropped. Thereafter, the discharge of articles into the receiver and the dropping of them into the lower receiver and finally the receptacle **80** continues as described above until a total of 31 has been dropped. Thus, no matter where the conveyor is located at the beginning of a load, the system will operate until a load of 31 has been dropped.

I claim:

1. A collator for articles comprising,
- a conveyor pivoted at its upstream end to swing its downstream end to any of a number of discharge points,
- means for swinging said conveyor to said discharge points to drop one of said articles at each of said discharge points,
- an upper receiver at said discharge end to receive articles from said conveyor,
- a plurality of spaced parallel ribs in said upper receiver forming a plurality of compartments for the receipt of articles at said discharge points,
- a lower receiver below said upper receiver for receiving a plurality of groups of articles from said upper receiver,
- trap doors forming the bottom walls of said receivers,
- and means for selectively opening said trap doors to first collect and drop a layer of articles from said upper receiver and to collect and drop a plurality of articles from said lower receiver.

2. A collator for articles as in claim **1** further comprising,
- an article detector at the downstream end of said conveyor,
- control means connected to said detector to count said articles and shift said conveyor from one discharge point to another as required to deposit a single layer of articles into said upper receiver.

3. A collator for articles as in claim **2** in which said detector includes a photoelectric cell.

4. A collator as in claim **2** in which said control means includes an adjustable counter for counting the total articles to be dropped into a receptacle, said counter causing the opening of said trap doors.

5. A collator as in claim **4** further comprising an adjustable counter for counting the articles per layer deposited in said upper receiver, said articles per layer counter causing the opening of said trap door in the upper receiver when the preselected number of articles is counted.

6. A collator for articles comprising,
- a conveyor pivoted at its upstream end to swing its downstream end to any of a number of discharge points,
- means for swinging said conveyor to said discharge points to drop one of said articles at each of said discharge points,
- an upper receiver at said discharge end to receive articles from said conveyor,
- a lower receiver below said upper receiver for receiving a plurality of groups of articles from said upper receiver,
- trap doors forming the bottom walls of said receivers,
- means for selectively opening said trap doors to first collect and drop a layer of articles from said upper receiver and to collect and drop a plurality of articles from said lower receiver,
- an article detector at the downstream end of said conveyor,
- an adjustable counter connected to said detector for counting the total articles to be dropped into a receptacle, said counter causing the opening of said trap doors,
- an adjustable counter for counting the articles per layer deposited in said upper receiver, said articles per layer counter causing the opening of said trap door at the upper receiver when the preselected number of articles is counted,
- limit switches on each side of said conveyor,
- said limit switches being operable to reset said articles per layer counter to zero whereby to permit said upper receiver to receive a full layer and a partial layer before its trap door opens.

7. A collator for articles comprising,
- a conveyor pivoted at its upstream end to swing its downstream end to any of a number of discharge points,
- means for swinging said conveyor to said discharge points to drop one of said articles at each of said discharge points,
- an upper receiver at said discharge end to receive articles from said conveyor,
- a lower receiver below said upper receiver for receiving a plurality of groups of articles from said upper receiver,
- trap doors forming the bottom walls of said receivers,
- means for selectively opening said trap doors to first collect and drop a layer of articles from said upper receiver and to collect and dorp a plurality of articles from said lower receiver,

an article detector at the downstream end of said conveyor,
an adjustable counter connected to said detector for counting the total articles to be dropped into a receptacle, said counter causing the opening of said trap doors,
an adjustable counter for counting the articles per layer deposited in said upper receiver, said articles per layer counter causing the opening of said trap door at the upper receiver when the preselected number of articles is counted,
an adjustable counter for counting the layers dropped into said lower receiver and operable to cause said trap door of said lower receiver to operate when a preselected number of layers is counted.

8. A collator for articles comprising,
a conveyor pivoted at its upstream end to swing its downstream end to any of a number of discharge points,
means for swinging said conveyor to said discharge points to drop one of said articles at each of said discharge points,
a receiver for articles located at the downstream end of said conveyor,
a trap door forming the bottom wall of said receiver,
an article detector at the downstream end of said conveyor,
control means connected to said detector to count said articles and shift said conveyor from one discharge point to another as each article passes the discharge end of said conveyor,
and a plurality of spaced parallel ribs on the upper surface of said trap door extending generally in the direction of said conveyor and forming compartments at each of said discharge points for the receipt of said articles.

9. A collator as in claim **8** in which said ribs are spaced apart a distance less than the width of the articles, thereby causing said articles to be deposited on said trap door in shingled relation.

5 10 15 20 25 30 35 40 45 50 55 60 65

10. A collator as in claim **8** and means for stopping said conveyor close to the leading rib of each compartment to cause articles to be shingled in two different orientations dependent upon the direction of movement of said conveyor.

11. A collator for articles comprising,
a conveyor pivoted at its upstream end to swing its downstream end to any of a number of discharge points,
means for swinging said conveyor to said discharge points to drop one of said articles at each of said discharge points,
a receiver for articles located at the downstream end of said conveyor,
a trap door forming the bottom wall of said receiver,
an article detector at the downstream end of said conveyor,
control means connected to said detector to count said articles and shift said conveyor from one discharge point to another as each article passes the discharge end of said conveyor,
said conveyor swinging means comprising,
a first fluid-operated double-acting piston and cylinder connected to said conveyor to cause said conveyor to swing,
a second fluid-operated double-acting piston and cylinder connected to said conveyor to brake said conveyor, and
means for operating said pistons and cylinders.

12. A collator as in claim **11** in which said operating means includes an electric eye mounted on said conveyor,
a series of fixed spaced indicia readable by said electric eye, each of said indicia corresponding to a discharge point,
and means connecting said electric eye to said second piston and cylinder to brake said conveyor at each discharge point.

* * * * *

Appendix C

October 1975

CLASS 3, ARTIFICIAL BODY MEMBERS

Original Classification 1915

1	MISCELLANEOUS
1.1	ELECTRICAL ACTUATION
1.2	FLUID ACTUATION
1.3	LARYNXES
1.4	ARTERY
1.5	HEART VALVE
1.7	ARTIFICIAL HEART
2	LEGS
4	. Extensions
5	. . Foot
14	. Torso actuated or controlled
15	. Torso attachments
16	. Suspenders or attachments from natural legs
17R	. Sockets
17SS	. . . Suction sockets
18	. . Yieldably mounted
19	. . Pads or liners
20	. . . Fluid
21	. Adjustable shank or thigh
22	. Knee
23	. . With foot actuator
24	. . . Latch
25	. . . Spring
26	. . Brakes and latches
27	. . . Position or weight responsive
28	. . . Adjustable friction joints
29	. . Springs
30	. Ankle
31	. . Universal joint
32	. . . Resilient
33	. . Resiliently actuated or controlled
34	. . . Elastic cords
35	. . . Springs
6	. Feet
6.1	. . Toe
7	. . Resilient
8	. . . Fluid
12	ARMS AND HANDS
12.1	. Torso supported and actuated
12.2	. Elbow
12.3	. . With forearm actuation
12.4	. Wrist
12.5	. . With wrist actuation
12.6	. Arm or torso initiated finger actuation
12.7	. Finger actuator embodied in simulated hand
12.8	. With article or article holder
1.9	BONE
1.91	. Joint
1.911	. . Knee
1.912	. . Hip
1.913	. . . Femoral head
13	EYES
36	BREASTS

Appendix D

No 3,225,299

THE UNITED STATES OF AMERICA

TO ALL TO WHOM THESE PRESENTS SHALL COME:

Whereas William H. Middendorf, of Covington, Kentucky,

PRESENTED TO THE Commissioner of Patents A PETITION PRAYING FOR THE GRANT OF LETTERS PATENT FOR AN ALLEGED NEW AND USEFUL INVENTION THE TITLE AND A DESCRIPTION OF WHICH ARE CONTAINED IN THE SPECIFICATION OF WHICH A COPY IS HEREUNTO ANNEXED AND MADE A PART HEREOF, AND COMPLIED WITH THE VARIOUS REQUIREMENTS OF LAW IN SUCH CASES MADE AND PROVIDED, AND

Whereas UPON DUE EXAMINATION MADE THE SAID CLAIMANT is ADJUDGED TO BE JUSTLY ENTITLED TO A PATENT UNDER THE LAW.

NOW THEREFORE THESE Letters Patent ARE TO GRANT UNTO THE SAID William H. Middendorf, his heirs OR ASSIGNS ...OR THE TERM OF SEVENTEEN YEARS FROM THE DATE OF THIS GRANT

...RIGHT TO EXCLUDE OTHERS FROM MAKING, USING OR SELLING THE SAID INVEN-... ...HROUGHOUT THE UNITED STATES.

In testimony whereof I have hereunto set my hand and caused the seal of the Patent Office to be affixed at the City of Washington this twenty-first day of December, in the year of our Lord one thousand nine hundred and sixty-five, and of the Independence of the United States of America the one hundred and ninetieth.

Attest:

Attesting Officer.

Commissioner of Patents.

FORM PO 377A (9-23-52)

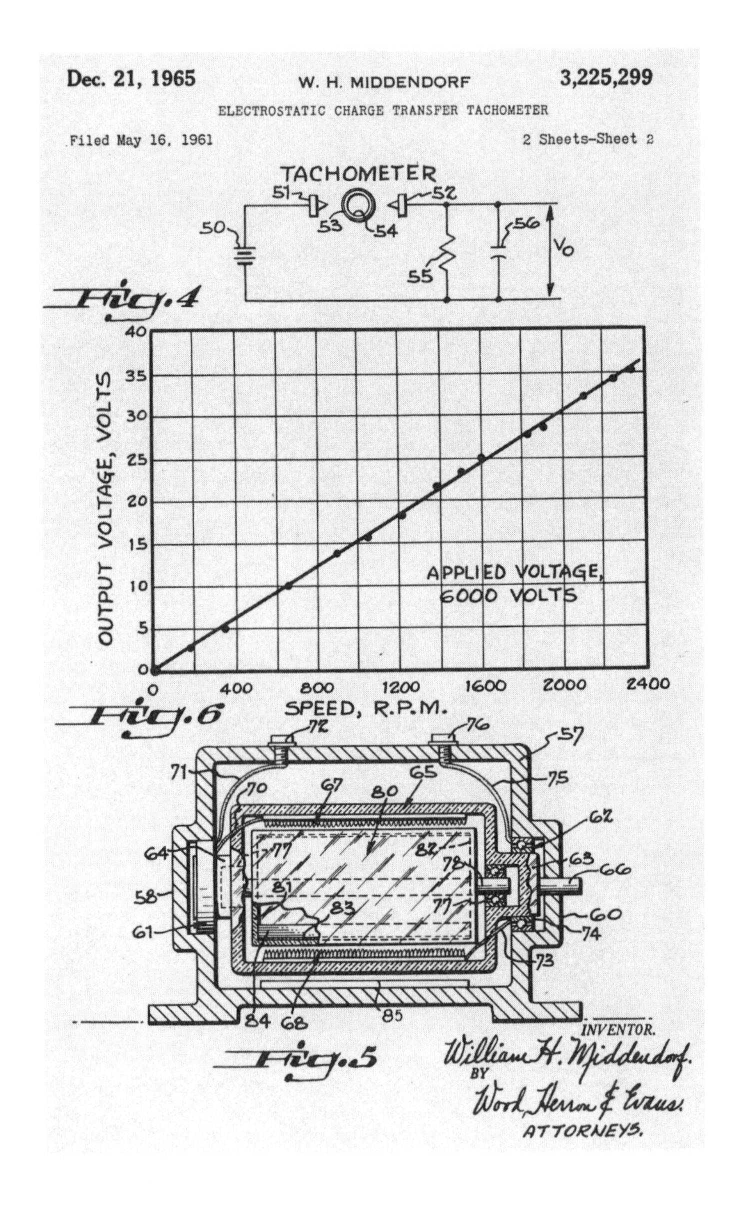

Dec. 21, 1965
W. H. MIDDENDORF
3,225,299
ELECTROSTATIC CHARGE TRANSFER TACHOMETER
Filed May 16, 1961
2 Sheets-Sheet 2
TACHOMETER
Fig. 4
OUTPUT VOLTAGE, VOLTS
SPEED, R.P.M.
APPLIED VOLTAGE, 6000 VOLTS
Fig. 6
Fig. 5
INVENTOR.
William H. Middendorf
BY
Wood, Herron & Evans
ATTORNEYS.

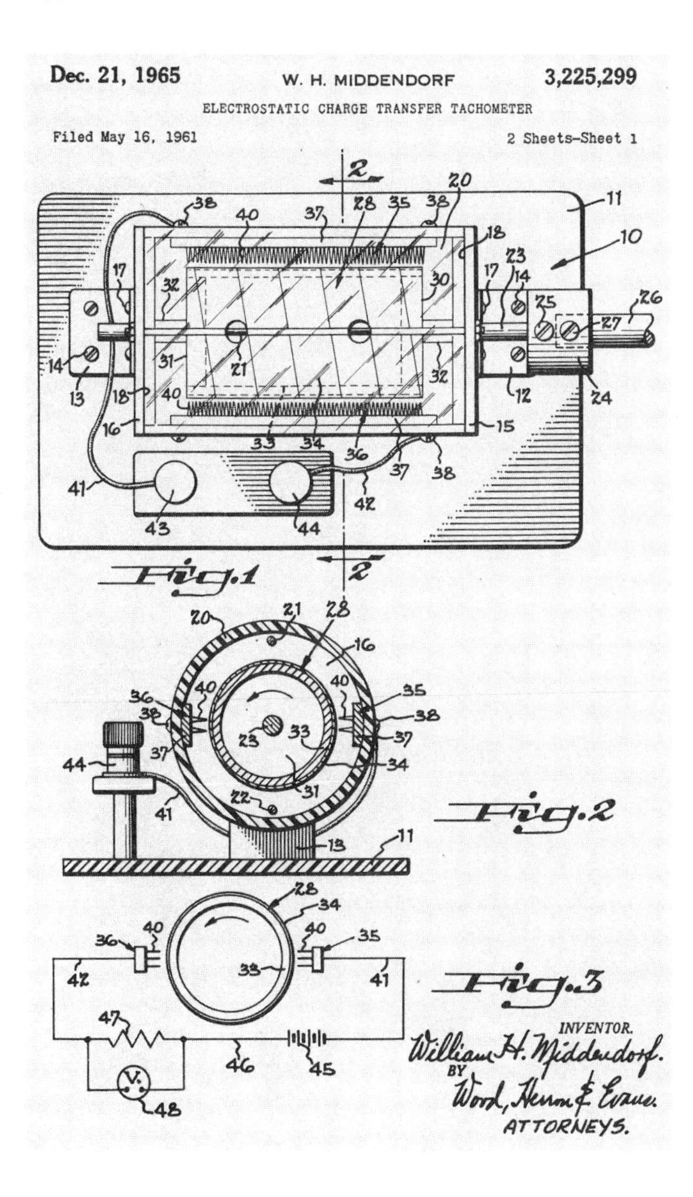
Dec. 21, 1965
W. H. MIDDENDORF
3,225,299
ELECTROSTATIC CHARGE TRANSFER TACHOMETER
Filed May 16, 1961
2 Sheets-Sheet 1
Fig.1
Fig.2
Fig.3
INVENTOR.
William H. Middendorf.
BY
Wood, Herron & Evans.
ATTORNEYS.

United States Patent Office 3,225,299
Patented Dec. 21, 1965

1

3,225,299
ELECTROSTATIC CHARGE TRANSFER TACHOMETER
William H. Middendorf, 407 Kentucky Drive, Fort Wright, Covington, Ky.
Filed May 16, 1961, Ser. No. 110,570
13 Claims. (Cl. 324—70)

The present invention relates to measuring devices and is particularly directed to a novel tachometer for measuring the speed of a rotating shaft.

In the past, many different types of tachometers have been proposed for measuring the speed of a rotating body. While these devices have proven satisfactory for many uses, each of the devices has inherent disadvantages. For example, it has previously been suggested to measure shaft speed by means of miniature D.C. or A.C. induction tachometer generators. A D.C. generator is disadvantageous since the output voltage of such a generator contains a low frequency ripple voltage due to the fact that the voltage is rectified by means of commutator bars and brushes. Also, at high velocities, brush jump occurs which causes spurious voltages of substantial magnitude. Moreover, the brushes required in a D.C. generator produce an appreciable amount of friction, i.e., of the order of 1 oz.-inch. When a D.C. tachometer generator is used to measure the speed of a small, low power motor the added frictional load at times adversely affects the operation of devices driven by the motor. Also, the rotor possesses a high inertia. Thus, it is difficult to utilize a D.C. generator to accurately measure transient speed changes of a small motor.

The A.C. drag cup induction generators which have been proposed for use as tachometers eliminate certain disadvantages of the D.C. generator, such as brush friction and high inertia. However, the A.C. drag cup generators also suffer from certain inherent drawbacks. Among these drawbacks is the fact that the drag cup generator output contains third and fifth harmonic components caused by the non-linearity of the magnetic circuit. These generators are also disadvantageous because they produce a residual voltage, i.e., the voltage output is not zero for zero shaft speed.

A still further disadvantage of A.C. generators is that the phase of the output signal is shifted whenever a speed change occurs in the generator shaft. Thus, if the shaft speed changes rapidly there is an inherent time delay in the output signal. Still a further disadvantage of these generators is that they required the use of additional rectifying and filtering circuit components in order to provide a direct current output, such as is frequently required in servo mechanism controls.

Another type of tachometer which has previously been suggested in the pulse counting type. In one form of tachometer of this type light pulses are reflected from light marks provided on the rotating shaft, the speed of which is to be measured. In order modifications, slits provided in a thin disc are effective to pass a number of light pulses proportional to the speed of the rotating shaft. However, no matter what type of mechanism is employed to produce the pulse train, it will be appreciated that in any event the signal originally obtained is in the form of a train of frequency modulated pulses. A substantial amount of expensive and relatively complex circuit components are required to transform this frequency modulated pulse train to a D.C. voltage, the magnitude of which is correlated with shaft speed.

2

The principal object of the present invention is to provide a novel tachometer of exceedingly simple construction which is free from the defects inherent in the prior art devices and which combines for the first time the desiderata of a liner direct current output, no residual output, an exceedingly rapid electrical response, low inertia and low friction.

More particularly, the present invention is predicated upon the concept of utilizing the rate of transfer of charges, electrostatically deposited upon the dielectric surface of a rotor, from one electrode to another as a measure of velocity. Specifically, the present invention comprehends a tachometer including two stationary electrodes disposed in spaced relationship to a rotatable member interconnected to the shaft of unknown speed. This member includes a conductive portion and a dielectric or insulated sheath. A sufficiently high D.C. potential is applied across the electrodes to provide an elecrostatic field effective to cause field emission to occur from the negative electrode. Electrons emitted from this electrode move in the presence of the field toward the conductive portion of the rotor and in the process of their movement become trapped upon the area of the dielectric sheath immediately adjacent to the emitting electrode.

As the rotor is moved in conformity with the rotation of the shaft of unknown speed, the charged area of the rotor is shifted to a position adjacent to the second or positive electrode. Under the influence of the field adjacent this electrode, the electrons previously stored upon the rotor are drawn to the positive electrode which is effect wipes the rotor clean. These electrons are returned from the positive electrode to the power source.

In accordance with the present invention, a resistor is placed in electrical series connection with the electrodes and power source and the current flowing in the circuit is measured by measuring a potential developed across this resistor. For a given interelectrode potential, the current flow in the circuit connecting the electrodes varies with the rate of which charges are carried by the rotor from the negative electrode to the positive electrode. I have determined that this current varies linearly with the speed of rotation of the rotor shaft, and that a zero current flows at zero shaft speed.

One of the principal advantages of the present tachometer is that it is of extremely simple construction. Moreover, the output circuit for obtaining a reading of the tachometer output is equally simple there being no need for special rectifiers, frequency responsive circuits, or the like.

Another advantage of the present invention is that the output signal of the tachometer is already in the most useful form for use in conjunction with servo mechanism units and the like. More particularly, the present tachometer unit produces an amplitude modulated direct current signal having no residual component so that the output signal is zero for zero output shaft speed. Moreover, the output signal of the present tachometer varies linearly with shaft speed.

Still another advantage of the output signal of the present tachometer is that it varies instantaneously with changes in shaft speed.

A further advantage of the present tachometer is that the device interferes only to a minimum extent with the normal operation of the device being measured even though that device may be driven by a relatively low powered motor or the like. Thus, the present tachometer requires no commutators, brushes or other frictional contact elements. Consequently, the tachometer introduces only the minimum friction inherent in the rotor shaft bearings. Moreover, the rotor has a very low inertia since it comprises only a conductive portion and an insulating sheath both of which can be formed from extremely lightweight materials. Thus, for example, in one embodiment the rotor may be formed of a glass cylinder, the interior of which is coated with a conductive paint. In this embodiment, the paint constitutes the conductive portion of the rotor while the glass cylinder constitutes the dielectric sheath.

These and other objectives and advantages of the present invention will be more readily apparent from a consideration of the following detailed description of the drawings illustrating a preferred embodiment of the invention.

In the drawings:

FIGURE 1 is a top plan view of a tachometer constructed in accordance with the principles of the present invention.

FIGURE 2 is a cross sectional view of the tachometer taken along line 2—2 of FIGURE 1.

FIGURE 3 is a schematic circuit diagram of the tachometer shown in FIGURE 1.

FIGURE 4 is a schematic circuit diagram of a slightly modified form of circuit including the present tachometer.

FIGURE 5 is a longitudinal cross sectional view partially broken away of a modified form of tachometer.

FIGURE 6 is a graph showing the relationship of the voltage output of the tachometer with the speed of the tachometer rotor.

As is shown in FIGURE 1, the tachometer **10** comprises a base **11** which supports two spaced angle bracket members **12** and **13**. These members are secured to the base in any suitable manner, such as by means of bolts **14**. Each of the angle brackets supports an end plate **15** and **16**. End plates **15** and **16** are formed of any suitable material, such as aluminum, and are secured to the angle brackets in any suitable manner, such as by means of bolts **17**.

As shown in FIGURES 1 and 2, each of the end plates **15** and **16** is provided with an annular shoulder **18** which receives the end of a housing cylinder **20**. Housing cylinder **20** is formed of any suitable electrical insulating material, such as plexiglass. Two horizontal tie rods **21** and **22** interconnect the opposite end plates **15** and **16** to hold those plates tightly against cylinder **20**. The cylinder **20** is thus mounted rigidly to the end plates and cannot rotate relative thereto.

An aperture is formed in the center of each of the end plates **15** and **16** and suitable ball bearings (not shown) are mounted in these apertures. A tachometer shaft **23** extends axially through cylinder **20** and is rotatably journalled in the two end roller bearings. One end of tachometer shaft **23** carries a coupling sleeve **24** which is held fast upon the shaft by means of a set screw **25**. Coupling sleeve **24** is adapted to receive a shaft **26**, the speed of which it is desired to measure. Sleeve **24** is secured to shaft **26** in any suitable manner, such as by means of a set screw **27**.

Tachometer shaft **23** carries cylindrical drum member **28**. Cylinder **28** comprises two end members **30** and **31** formed of any suitable lightweight material, for example aluminum stock. Each of these end members **30** and **31** is provided with a central bore which is press fit over shaft **23**. Shaft **23** and hence drum **28** are held against longitudinal movement relative to cylinder **20** by means of bushings **32** which are disposed between end members **30** and **31** of the drum and the shaft bearings provided in the end plates **15** and **16**. An electrically conductive cylinder **33** is carried by end members **30** and **31**. This cylinder is formed of a lightweight conductive material such as aluminum having a wall thickness of 1/16 of an inch. The cylinder is press fit over end members **30** and **31**. The outer surface of conductive cylinder **33** is covered with a suitable dielectric sheath **34**. One satisfactory form of sheath comprises a silicone tape wrapped around the periphery of the cylinder. In addition to the tape, the ends of the cylinder are painted with a suitable electrical insulating material, such as Formvar, to prevent arcing around the ends of the tape.

In addition to the rotor **28**, housing cylinder **20** encloses two electrodes, or charging combs **35** and **36**. Each of the charging combs comprises a conductive plate **37** formed of brass, or the like, secured as by means of bolts **38** to the inner surface of cylindrical housing **20**. Each of these plates extends longitudinally of the housing and preferably the plates are disposed at diametrically opposite positions on the housing wall.

Each of the plates **37** carries a plurality of inwardly facing conductive needles **40**. In the specific embodiment shown, the needles are formed of two rows of steel phonograph needles placed on 3/32 inch centers and press fit into suitable openings drilled in the brass support plates. The air gap between the points of the needles and the dielectric sheath, in the embodiment shown, is substantially 1/16 of an inch, although as shorter air gap is to be preferred. Conductive leads **41** and **42** are joined to each of the brass plates **37** and are respectively connected to output terminals **43** and **44** of the tachometer.

In addition to these elements, the present tachometer comprises a regulated D.C. power supply indicated diagrammatically at **45**. The negative terminal of this power supply is connected to a negatively charged comb, or electrode, **35**. The positive terminal of power supply **45** is connected through a lead **46** to one side of a load resistor **47**, the other side of the resistor being connected to conductor **42** which is in turn joined to positive electrode **36**. The output signal of the tachometer is measured by a suitable voltmeter **48** connected across resistor **47**.

In operation, sleeve **24** is coupled to a shaft **26**, the speed of which is to be determined. D.C. power supply **45** is energized to apply to electrodes **35** and **36** a sufficiently high voltage to cause a high field emission between them. In the embodiment shown, the potential applied to the electrodes **35** and **36** is of the order of 6000 volts. It will, of course, be appreciated that the requisite potential depends upon the electrode spacing, electrode temperature, and electrode construction. For example, a substantially lower potential is required if special electrodes are used. For example, if tungsten needle points are employed, such as those described in a paper entitled IRE Trans. on Military Electronics, volume MIL-4, No. 1, January 1960, pages 38–45, a voltage of only 320 volts between the electrodes **35** and **36** is requisite. Also, if the negative electrode is heated to provide thermionic emission, the necessary voltage is still further reduced to that necessary to attract the free electrons to the positive electrode **36** pulse the drop across the load resistor **47**.

In any event, when a sufficient potential is applied across electrodes **35** and **36**, electrons leave the negative electrode and are drawn toward the insulated conductive rotor cylinder **33**. These electrons are trapped, however, on the outer surface of the dielectric sheath. Consequently, these trapped electrons tend to prevent other electrons from leaving the electrode points. In fact, if the rotor is stationary, the flow of electrons from the negative electrode to the surface of the sheath rapidly diminishes to zero and the unit merely functions as a simple capacitor of relatively low capacitance.

However, when shaft **26** is rotated, it causes a similar rotation in shaft **23**. The area of the sheath **34** covered with electrons moves away from the negative electrode comb **35** toward the positive electrode comb **36**. As drum **28** rotates, other electrons are emitted from comb **35**, and these electrons flow to the newly exposed, uncharged areas of the rotor then adjacent to the electrode points of comb **35**.

At the same time that areas of the dielectric sheath are being charged by electrons emitted from negative electrode **35**, other electrons are being drawn from the surface of the dielectric sheath in the area immediately adjacent to positive electrode **36**. This electrode is effective to attract all of the electrons previously stored in are area of the drum now adjacent to the positive electrode and thus in effect wipes the rotor clean. The electrons attracted to the positive electrode **36** flow through conductor **42** and load resistor **47** and are returned to the positive terminal of the D.C. power supply **45**. Since resistor **47** is placed in series circuit relation with the power supply and electrodes **35** and **36**, the rate of flow of electrons, or current, is readily measured by means of a suitable voltage measuring device connected across resistor **47**.

It has been empirically determined that the voltage output of the tachometer varies linearly with the speed of the tachometer shaft and hence with the shaft being measured. FIGURE 6 is a graph in which the output voltage is plotted against the shaft speed in r.p.m.'s. It will be noted that the output voltage measured across resistor **47** varies linearly (within the limits of experimental error) from zero volts at zero r.p.m.'s to 35 volts at approximately 2350 r.p.m.'s.

A theoretical explanation of this linear relationship can be obtained by analogizing the tachometer to a diode vacuum tube in which the plate current is inhibited by the insulating sheath which is placed between the anode, or positive electrode, **36** and the cathode, or negative electrode **35**. Since the time required to saturate an area of the insulated cylinder, or sheath with charge is negligible compared to the time necessary to move the section from the influence of the negative electrode **35**, the charge of the unit area will be a function of voltage and spacing. Specifically,

$$\frac{dq}{dA}=f(v, l)$$

Where dA is the differential area charged, v is the applied voltage and l is a characteristic distance.

But

$$i\frac{dq}{dt}=\frac{dq}{dA}\frac{dA}{dt}$$

and $dA=ard\theta$ if r is the outer radius of the rotor, and a is its axial length. This gives

$$i=f(v, l)ar\frac{d\theta}{dt}$$

Hence

$$i=[f(v,l)ar]\omega$$

The last equation shows that so long as the voltage applied across electrodes **35** and **36** by power source **45** is constant, the tachometer functions as a current generator, the output of which is directly proportioned to angular speed ω. The results predicted by this equation conform exactly with the empirical current speed relationships plotted in FIGURE 6.

It will be appreciated that so long as the velocity of the shaft **23**, and hence the velocity of the surface of cylinder **28**, is small in comparison with the speed of charge transfer between the electrodes and cylinder, there is no time lapse between any change in the speed of shaft **26** and the rate of change of voltage sensed by voltmeter **48**. However, as a practical matter, in the embodiment shown in FIGURE 1, the output signal sensed by voltmeter **48** contains a large noise component as is to be expected in devices having electric discharge through

gas. This noise component presents no difficulty if an averaging device, such as a vacuum tube voltmeter is utilized to indicate steady state speed. If, however, transient response is desired this noise component is troublesome.

One compromise solution to obtaining a transient response with an acceptable noise level is indicated in the circuit diagram of FIGURE 4. In that circuit, a D.C. power supply **50** is connected in series with positive electrode **51** and negative electrode **52** of the tachometer. These electrodes are spaced relative to a rotating insulated sheath **53** surrounding a conductive cylinder **54** in exactly the same way as is shown in FIGURES 1 and 2. A load resistor **55** is also placed in series with the electrodes and D.C. power supply. However, in the embodiment shown in FIGURE 4, a capacitance **56** is connected in parallel with the load resistance and the output voltage indicated by the symbol V_o is measured by a suitable device such as an oscilloscope connected across this parallel combination. The bypass capacitor functions to remove a portion of the high frequency noise. However, this capacitor does introduce into the circuit a time constant which limits the speed of response of the tachometer.

I have determined that in order to obtain optimum results from the tachometer, i.e., instantaneous transient response with no noise appearing in the output signal, the space surrounding the rotor and intermediate the rotor and electrodes should be evacuated. Obviously, if the unit disclosed in FIGURE 1 is employed in outer space, this vacuum condition is an ambient condition and no special evacuation of housing **20** is required.

On the other hand, where a vacuum is not present as an ambient condition, a modified form of tachometer, such as that shown in FIGURE 5, is highly advantageous. As is there shown, the modified form of tachometer includes an outer housing **57** formed of metal, or any other suitable material. This housing is provided with two end bells **58** and **60** which house ball bearings **61** and **62**. These ball bearings rotatably support hub portions **63** and **64** formed on inner housing **65**. The inner housing is made of any light weight non-conductive material, such as glass.

Hub **63** of the inner housing is secured to a metal shaft **66** in any suitable manner. This shaft extends outwardly through an aperture in end bell **60**. It is to be understood that the outer end of shaft **66** carries a suitable coupling member (not shown), such as a sleeve, similar to sleeve **24** for joining shaft **66** to a shaft, the speed of which is to be determined. It will be appreciated that inner housing **65** constitutes an integral closed chamber which is evacuated in any suitable manner. Inner housing **65** supports positive and negative electrode combs **67** and **68**. These combs are identical with the combs shown in FIGURES 1 and 2 and are preferably disposed at diametrically opposed positions on the inner housing walls. One electrical lead **70** is taken from electrode **67** and is passed through glass housing **65** in any suitable manner, such as a glass to Kovar seal.

Lead **70** is connected to the inner race of ball bearing **61**, while a second lead **71** is connected to the outer race of the ball bearing and to an output terminal **72**. In a similar manner, electrode **68** is joined through a lead **73** passing through glass housing **65** to the inner race **74** of ball bearing **62**, while a second lead **75** joins the outer race of this ball bearing with terminal **76**.

These last named electrical connections facilitate the application of a D.C. potential across electrodes **67** and **68** despite the fact that in this modified embodiment the electrodes are rotated in synchronism with the shaft, of unknown velocity.

As is shown in FIGURE 5, each of the hub members **63** and **64** of inner housing **65** is hollow to provide a seat for an inner ball bearing **77—77**. These ball bearings rotatably support a drum shaft **78**. This drum shaft supports a cylindrical drum. In this embodiment, the drum comprises a cylindrical glass shell **80** wihch is fixedly mounted upon shaft **78** by means of suitable end members **81** and **82**. The inner surface of glass cylinder **80** is coated with a conductive coating **83**. It will be appreciated that in this embodiment, the paint coating **83** comprises a conductive cylinder which is functionally the same as the aluminum cylinder **28** in FIGURES 1 and 2, while the glass cylinder **80** is the functional equivalent of the insulating sheath formed by tape **34**.

In the modified embodiment of FIGURE 5, means are also provided for holding the drum **80** stationary. One suitable form of such means comprises an elongated weight **84** mounted on the bottom of cylinder **80**. The inertia of this member is sufficient to hold the cylinder substantially stationary against the frictional force imposed by inner bearings **77—77**. Alternatively, weight **84** can be formed of a ferro magnetic material which cooperates with a permanent magnet, such as magnet **85**, mounted upon stationary housing **57**.

In the embodiment shown in FIGURE 5, a potential is applied across electrodes **67** and **68** in the same manner as in the embodiment of FIGURES 1 and 2. In the modified embodiment, electrons are drawn from the negative electrode by high field emission and collect upon the adjacent area of the insulated glass cylinder **80**. As the positive electrode sweeps over the areas previously charged by the negative electrode, the stored electrons are attracted to the positive electrode and flow through a circuit including a load resistor to the power supply. Again, the current flowing through the circuit varies linearly with the velocity of shaft **66**. Also, there is no current output for zero rotational speed of shaft **66**.

From the foregoing disclosure of the general principles of the present invention and the above description of two preferred embodiments, those skilled in the art will readily comprehend the various modifications to which the present invention is susceptible. Thus, for example, it is contemplated that electrons can be drawn from the cathode to the rotor and can be drawn from the rotor to the anode in the presence of a magnet, as opposed to an electrostatic field. It is further contemplated that the housing surrounding the electrodes and rotor in FIGURE 1 can be evacuated in such a modification. The rotor shaft in such a modification is terminated internally of the housing and a magnet is mounted on the end of this rotor shaft adjacent to the end wall of the housing. A similar magnet is disposed adjacent to the end wall of the housing externally of the housing for rotation co-axially with the inner magnet. This outer magnet is carried by a rotatable shaft carrying a coupling, such as coupling **24**, for joining the shaft to the shaft to be measured. Accordingly, I desire to be limited only by the scope of the following claims.

Having described my invention, I claim:

1. Apparatus for measuring the speed of a rotating member, said apparatus comprising a tachometer, said tachometer including a rotor rotatable about an axis and having a cylindrical conductive surface and a dielectric sheath overlying said surface, means for connecting said rotor for rotational movement with said rotating member, a first electrode disposed closely adjacent to said rotor but spaced from the dielectric sheath thereof, and a second electrode disposed closely adjacent to the rotor but spaced from the dielectric sheath portion thereof, said second electrode being spaced from said first electrode circumferentially with respect to said rotor, said electrodes extending axially of said rotor, means for applying a D.C. potential across said electrodes, said first electrode being effective to emit electrons, said second electrode being effective to collect electrons previously deposited on said sheath by said first electrode, and means responsive to the current flowing between said electrodes for indicating the velocity of said rotating member.

2. Apparatus for measuring the speed of a rotating member, said apparatus comprising a tachometer, said

tachometer including a rotor rotatable about an axis and having a rotational conductive surface and a dielectric sheath overlying said surface, means for connecting said rotor for movement with said rotating member, a first electrode disposed closely adjacent to said rotor but spaced from the dielectric sheath thereof, and a second electrode disposed closely adjacent to the rotor but spaced from the dielectric sheath portion thereof, said second electrode being spaced from said first electrode circumferentially with respect to said rotor, said electrodes extending axially of said rotor, means for applying a D.C. potential across said electrodes, said potential being sufficiently great to cause field emission from said first electrode, said second electrode being effective to collect electrons previously deposited on said sheath by said first electrode, and means responsive to the current flowing between said electrodes for indicating the velocity of said rotating member.

3. Apparatus for measuring the speed of a rotating member, said apparatus comprising a tachometer, said tachometer including a cylindrical conductive surface and a dielectric sheath overlying said surface, a first electrode disposed closely adjacent to but spaced from said dielectric sheath, and a second electrode disposed closely adjacent to but spaced from said dielectric sheath, said second electrode being spaced from said first electrode circumferentially with respect to said cylindrical conductive surface, means for effecting rotating movement of said conductive surface and dielectric sheath relative to said electrodes in accordance with the speed of said rotating member, means for applying a D.C. potential across said electrodes, said first electrode being effective to emit electrons, said second electrode being effective to collect electrons previously deposited on said sheath by said first electrode, and evacuated housing means surrounding said electrodes, said conductive surface and said dielectric, and means responsive to the current flowing between said electrodes for indicating the velocity of said rotating member.

4. In apparatus for measuring the speed of a rotating member, a device comprising a housing, a shaft adapted to be interconnected to said member, means associated with said housing for rotatably journalling said shaft, a cylindrical rotor carried by said shaft, said cylindrical rotor comprising a conductive cylinder, and a dielectric cylinder surrounding said conductive cylinder, an emitter electrode closely spaced from the surface of said cylinder, a second electrode spaced from said cylinder in a position spaced from said first electrode circumferentially with respect to said rotor, said electrodes extending axially of said rotor, said first electrode being effective to emit electrons, said second electrode being effective to collect electrons previously deposited on said sheath by said first electrode.

5. In apparatus for measuring the speed of a rotating member, a device comprising a housing, a shaft adapted to be interconnected to said member, means carried by said housing for rotatably journalling said shaft, a cylindrical rotor carried by said shaft, said cylindrical rotor comprising a conductive cylinder coaxial with said shaft, and a dielectric sheath surrounding said conductive cylinder, a first electrode including a plurality of pointed electrode elements directed toward said dielectric sheath and being disposed adjacent to said sheath, and a second electrode including a plurality of pointed electrode elements directed toward and spaced from said sheath in an area spaced from said first electrode circumferentially with respect to said rotor, said electrodes extending axially of said rotor, said first electrode being effective to emit electrons, said second electrode being effective to collect electrons previously deposited on said sheath by said first electrode.

6. In apparatus for measuring the speed of a rotating member, a device comprising a housing, a shaft adapted to be interconnected to said member, means associated

with said housing for rotatably journalling said shaft, a cylindrical rotor carried by said shaft, said cylindrical rotor comprising a conductive cylinder and a dielectric sheath surrounding said conductive cylinder, a first electrode comb, said electrode comb comprising a plurality of needlelike electrode points directed toward said sheath and disposed in a band extending parallel to the axis of said rotor shaft, a second electrode comb assembly, means mounting said second electrode comb assembly within said housing, said second electrode comb assembly including a plurality of needlelike electrodes facing inwardly toward said rotor, said electrode elements being disposed in a band parallel to said rotor shaft, said first electrode comb and said second electrode comb being spaced circumferentially relative to said cylindrical rotor, said first electrode comb being effective to emit electrons, said second electrode comb being effective to collect electrons previously deposited on said dielectric sheath by said first electrode.

7. In apparatus for measuring the speed of a shaft, a device comprising an outer housing, an inner housing, means rotatably supporting said inner housing upon said outer housing, means mounted upon said inner housing for joining said housing to a shaft the speed of which is to be measured, said inner housing forming an evacuated chamber, a drum, means mounting said drum for rotation relative to said inner housing, two electrodes mounted within said chamber, said electrodes being spaced apart circumferentially with respect to said drum, said drum comprising a conductive cylinder and an insulating cylinder surrounding said conductive cylinder, and means for holding said drum stationary relative to said outer housing, one of said electrodes being effective to emit electrons and the other of said electrodes being effective to collect electrons previously deposited by the other of said electrodes on the insulating cylinder.

8. In apparatus for measuring the speed of a shaft, a device comprising an outer housing, an inner housing, means rotatably supporting said inner housing upon said outer housing, means mounted upon said inner housing for joining said housing to a shaft the speed of which is to be measured, said inner housing forming an evacuated chamber, a drum, means mounting said drum for rotation relative to said inner housing, two electrodes mounted within said chamber, said electrodes being spaced apart circumferentially with respect to said drum, said drum comprising a conductive cylinder and an insulating cylinder surrounding said conductive cylinder, and means for holding said drum stationary relative to said outer housing, said last named means comprising a weight secured to a wall of said drum, one of said electrodes being effective to emit electrons and the other of said electrodes being effective to collect electrons previously deposited by the other of said electrodes on the insulating cylinder.

9. In apparatus for measuring the speed of a shaft, a device comprising an outer housing, an inner housing, means rotatably supporting said inner housing upon said outer housing, means mounted upon said inner housing for joining said housing to a shaft the speed of which is to be measured, said inner housing forming an evacuated chamber, a drum, means mounting said drum for rotation relative to said inner housing, two electrodes mounted within said chamber, said electrodes being spaced apart circumferentially with respect to said drum, said drum comprising a conductive cylinder and an insulating cylinder surrounding said conductive cylinder, and means for holding said drum stationary relative to said outer housing, said last named means comprising a ferro-magnetic member carried by said drum and a cooperating magnet secured to said outer housing, one of said electrodes being effective to emit electrons and the other of said electrodes being effective to collect electrons previously deposited by the other of said electrodes on the insulating cylinder.

10. In apparatus for measuring the speed of a shaft, a device comprising an outer housing, an inner housing,

means rotatably supporting said inner housing upon said outer housing, means mounted upon said inner housing for joining said housing to a shaft the speed of which is to be measured, said inner housing forming an evacuated chamber, a drum, means mounting said drum for rotation relative to said inner housing, two electrodes mounted within said chamber, said electrodes being spaced apart circumferentially with respect to said drum, said drum comprising a conductive cylinder and an insulating cylinder surrounding said conductive cylinder, and means for holding said drum stationary relative to said outer housing, two electrical terminals mounted upon said outer housing, and means providing an electrical connection between said electrodes and terminals, said last named means comprising roller bearings having outer races and inner races, the outer races being secured to said outer housing and the inner races being secured to said inner housing, and comprising conductors extending from said electrodes respectively through the walls of said inner housing, said conductors being respectively secured to the inner races of each of said bearings, and second connectors joined respectively to the outer races of said bearings to said terminals, one of said electrodes being effective to emit electrons and the other of said electrodes being effective to collect electrons previously deposited by the other of said electrodes on the insulating cylinder.

11. In apparatus for measuring the speed of a rotating member, a device comprising a housing, a shaft, means carried by said housing for rotatably journalling said shaft, a cylindrical rotor carried by said shaft, said cylindrical rotor comprising a cylindrical glass tube and an electrically conductive coating formed on the inner surface of said tube, a first electrode including a plurality of pointed electrode elements directed toward said rotor and being disposed adjacent to said rotor, and a second electrode including a plurality of pointed electrode elements directed toward and spaced from said rotor in an area spaced from said first electrode circumferentially with respect to said rotor, said first electrode being effective to emit electrons and said second electrode being effective to collect electrons previously deposited on said glass tube by said first electrode.

12. In apparatus for measuring the speed of a rotating shaft, a device comprising an elongated anode, an elongated cathode, said cathode being disposed parallel to said anode, a rotating member adapted for connection to said shaft, said rotating member having a peripheral surface disposed adjacent to said anode and cathode and having an axis of rotation parallel thereto, said anode and cathode being spaced circumferentially relative to said rotating member, the peripheral surface of said rotating member being formed of a dielectric material, said rotating member having conductive portions disposed radially inwardly of said surface, whereby said dielectric material is interposed between said conductive portions and said anode and cathode, and means for connecting said rotatable member to a rotatable shaft, the speed of which is to be measured, said cathode being effective to emit electrons and said anode being effective to collect electrons previously deposited on said dielectric material by said cathode.

13. In apparatus for measuring the speed of a rotating shaft, a device comprising an elongated anode, an elongated cathode, said cathode being disposed parallel to said anode, a rotating member adapted for connection to said shaft, said rotating member disposed adjacent to said anode and cathode and having an axis of rotation parallel thereto, said anode and cathode being spaced circumferentially relative to said rotating member, said rotating member having a peripheral surface formed of dielectric material and having conductive portions disposed radially inwardly of said surface whereby said dielectric material is interposed between said conductive portions and said anode and cathode, and evacuated means surrounding said anode, cathode and rotor, and means for effecting

3,225,299

11

relative rotation between said rotor and said anode and cathode in accordance with the speed of rotation of said rotating shaft, said cathode being effective to emit electrons and said anode being effective to collect electrons previously deposited on said dielectric material by said cathode.

References Cited by the Examiner

UNITED STATES PATENTS

2,545,354	3/1951	Hansen	310—5
2,667,615	1/1954	Brown	317—250 X
2,831,988	4/1958	Morel	310—6
3,024,371	3/1962	Lefkowitz	310—5

12

5 WALTER L. CARLSON, *Primary Examiner.*

SAMUEL BERNSTEIN, *Examiner.*

Appendix E

United States Patent Office

925,164
Registered Dec. 7, 1971

PRINCIPAL REGISTER
Trademark

Ser. No. 371,981, filed Sept. 28, 1970

Totes Incorporated (Ohio corporation)
Loveland, Ohio 45140

For: UMBRELLAS, in CLASS 41 (INT. CL. 18).
First use July 29, 1970; in commerce July 29, 1970.
Owner of Reg. No. 524,181.

Appendix F

United States Patent [19]

Leder et al.

[11] **Patent Number:** **4,736,866**

[45] **Date of Patent:** **Apr. 12, 1988**

[54] **TRANSGENIC NON-HUMAN MAMMALS**

[75] Inventors: **Philip Leder,** Chestnut Hill, Mass.; **Timothy A. Stewart,** San Francisco, Calif.

[73] Assignee: **President and Fellows of Harvard College,** Cambridge, Mass.

[21] Appl. No.: **623,774**

[22] Filed: **Jun. 22, 1984**

[51] **Int. Cl.**[4] **C12N 1/00;** C12Q 1/68; C12N 15/00; C12N 5/00

[52] **U.S. Cl.** **800/1;** 435/6; 435/172.3; 435/240.1; 435/240.2; 435/320; 435/317.1; 935/32; 935/59; 935/70; 935/76; 935/111

[58] **Field of Search** 435/6, 172.3, 240, 317, 435/320, 240.1, 240.2; 935/70, 76, 59, 111, 32; 800/1

[56] **References Cited**

U.S. PATENT DOCUMENTS

4,535,058 8/1985 Weinberg et al. 435/91
4,579,821 4/1986 Palmiter et al. 435/240

OTHER PUBLICATIONS

Ucker et al, Cell 27:257–266, Dec. 1981.
Ellis et al, Nature 292:506–511, Aug. 1981.
Goldfarb et al, Nature 296:404–409, Apr. 1981.
Huang et al, Cell 27:245–255, Dec. 1981.
Blair et al, Science 212:941–943, 1981.
Der et al, Proc. Natl. Acad. Sci. USA 79:3637–3640, Jun. 1982.
Shih et al, Cell 29:161–169, 1982.
Gorman et al, Proc. Natl. Acad. Sci. USA 79:6777–6781, Nov. 1982.
Schwab et al, EPA–600/9–82–013, Sym: Carcinogen, Polynucl. Aromat. Hydrocarbons Mar. Environ., 212–32 (1982).
Wagner et al. (1981) Proc. Natl. Acad. Sci USA 78, 5016–5020.
Stewart et al. (1982) Science 217, 1046–8.
Costantini et al. (1981) Nature 294, 92–94.
Lacy et al. (1983) Cell 34, 343–358.
McKnight et al. (1983) Cell 34, 335.
Binster et al. (1983) Nature 306, 332–336.
Palmiter et al. (1982) Nature 300, 611–615.
Palmiter et al. (1983) Science 222, 814.
Palmiter et al. (1982) Cell 29, 701–710.

Primary Examiner—Alvin E. Tanenholtz
Attorney, Agent, or Firm—Paul T. Clark

[57] **ABSTRACT**

A transgenic non-human eukaryotic animal whose germ cells and somatic cells contain an activated oncogene sequence introduced into the animal, or an ancestor of the animal, at an embryonic stage.

12 Claims, 2 Drawing Sheets

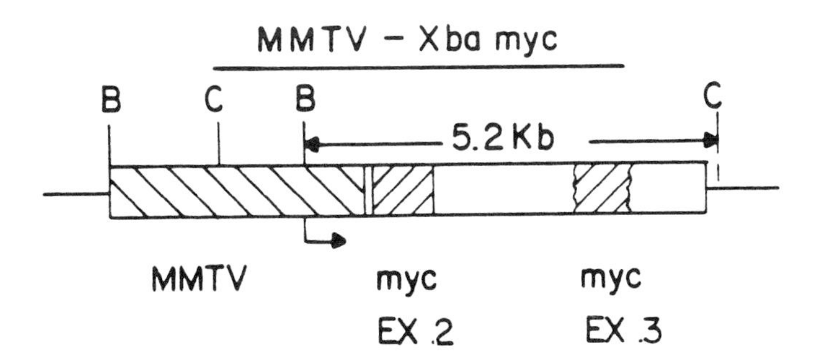

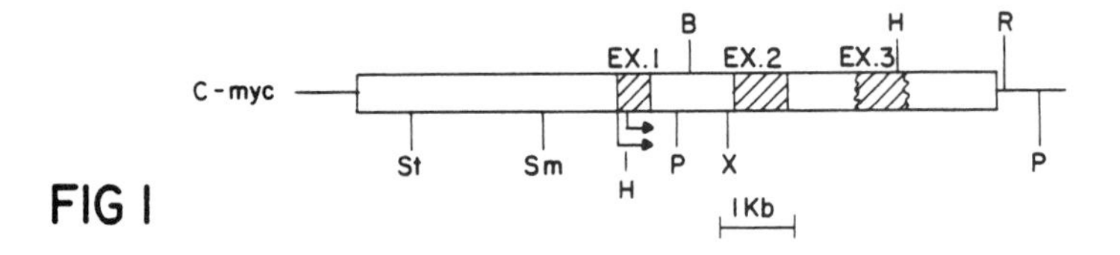

FIG 1

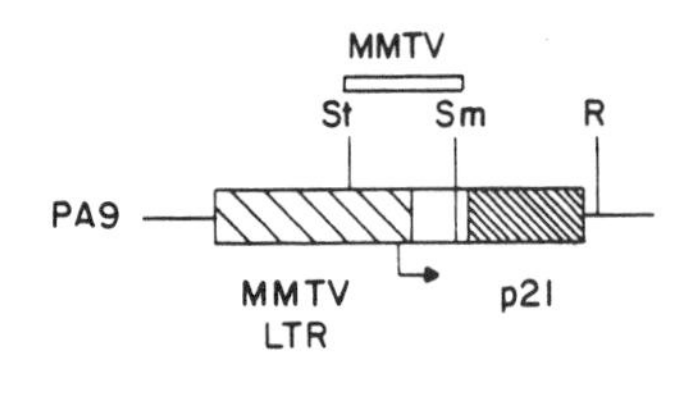

FIG 2

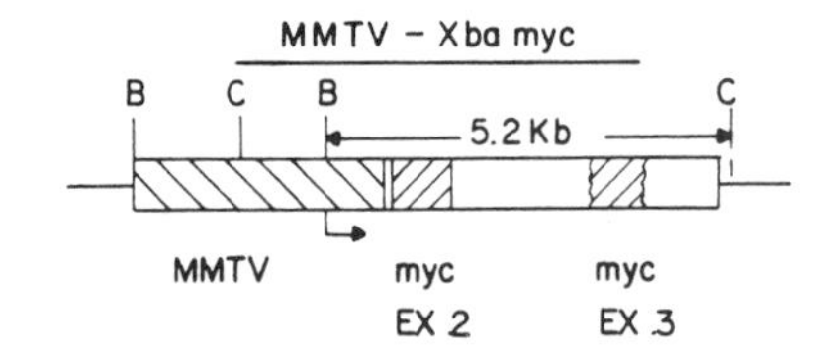

FIG 3

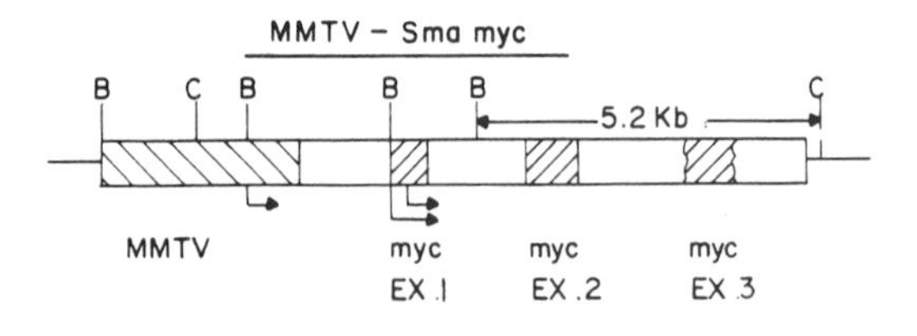

FIG 4

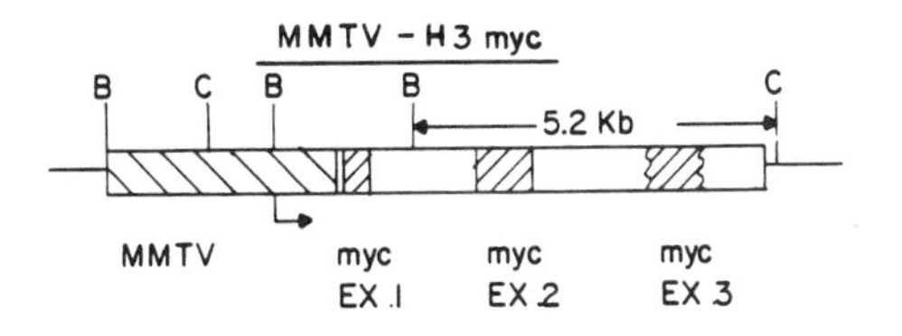

FIG 5

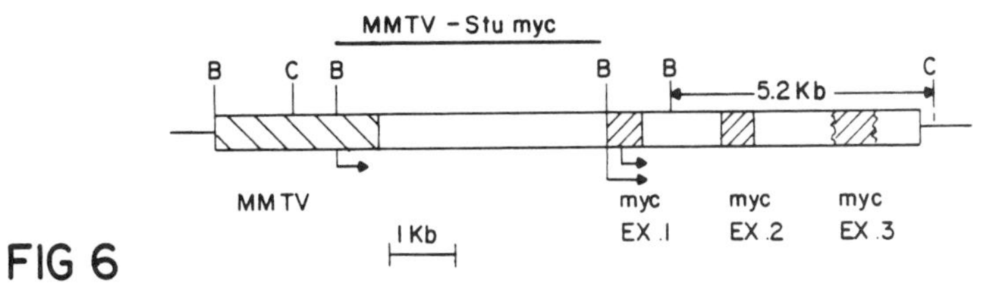

FIG 6

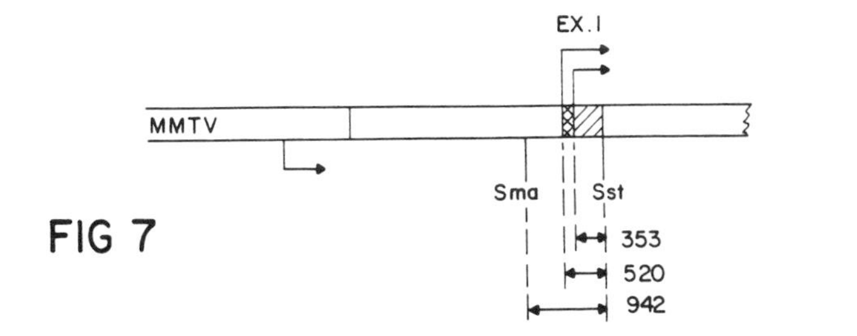

FIG 7

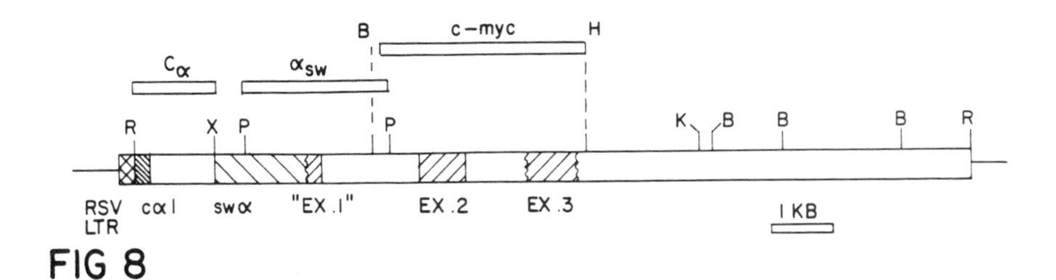

FIG 8

4,736,866

TRANSGENIC NON-HUMAN MAMMALS

BACKGROUND OF THE INVENTION

This invention relates to transgenic animals.

Transgenic animals carry a gene which has been introduced into the germline of the animal, or an ancestor of the animal, at an early (usually one-cell) developmental stage. Wagner et al. (1981) *P.N.A.S. U.S.A.* 78, 5016; and Stewart et al. (1982) *Science* 217, 1046 describe transgenic mice containing human globin genes. Constantini et al. (1981) *Nature* 294, 92; and Lacy et al. (1983) *Cell* 34, 343 describe transgenic mice containing rabbit globin genes. McKnight et al. (1983) *Cell* 34, 335 describes transgenic mice containing the chicken transferrin gene. Brinster et al. (1983) *Nature* 306, 332 describes transgenic mice containing a functionally rearranged immunoglobulin gene. Palmiter et al. (1982) *Nature* 300, 611 describes transgenic mice containing the rat growth hormone gene fused to a heavy metal-inducible metalothionein promoter sequence. Palmiter et al. (1982) *Cell* 29, 701 describes transgenic mice containing a thymidine kinase gene fused to a metalothionein promoter sequence. Palmiter et al. (1983) *Science* 222, 809 describes transgenic mice containing the human growth hormone gene fused to a metalothionein promoter sequence.

SUMMARY OF THE INVENTION

In general, the invention features a transgenic non-human eukaryotic animal (preferably a rodent such as a mouse) whose germ cells and somatic cells contain an activated oncogene sequence introduced into the animal, or an ancestor of the animal, at an embryonic stage (preferably the one-cell, or fertilized oocyte, stage, and generally not later than about the 8-cell stage). An activated oncogene sequence, as the term is used herein, means an oncogene which, when incorporated into the genome of the animal, increases the probability of the development of neoplasms (particularly malignant tumors) in the animal. There are several means by which an oncogene can be introduced into an animal embryo so as to be chromosomally incorporated in an activated state. One method is to transfect the embryo with the gene as it occurs naturally, and select transgenic animals in which the gene has integrated into the chromosome at a locus which results in activation. Other activation methods involve modifying the oncogene or its control sequences prior to introduction into the embryo. One such method is to transfect the embryo using a vector containing an already translocated oncogene. Other methods are to use an oncogene whose transcription is under the control of a synthetic or viral activating promoter, or to use an oncogene activated by one or more base pair substitutions, deletions, or additions.

In a preferred embodiment, the chromosome of the transgenic animal includes an endogenous coding sequence (most preferably the c-myc gene, hereinafter the myc gene), which is substantially the same as the oncogene sequence, and transcription of the oncogene sequence is under the control of a promoter sequence different from the promoter sequence controlling transcription of the endogenous coding sequence. The oncogene sequence can also be under the control of a synthetic promoter sequence. Preferably, the promoter sequence controlling transcription of the oncogene sequence is inducible.

Introduction of the oncogene sequence at the fertilized oocyte stage ensures that the oncogene sequence will be present in all of the germ cells and somatic cells of the transgenic animal. The presence of the onocogene sequence in the germ cells of the transgenic "founder" animal in turn means that all of the founder animal's descendants will carry the activated oncogene sequence in all of their germ cells and somatic cells. Introduction of the oncogene sequence at a later embryonic stage might result in the oncogene's absence from some somatic cells of the founder animal, but the descendants of such an animal that inherit the gene will carry the activated oncogene in all of their germ cells and somatic cells.

Any oncogene or effective sequence thereof can be used to produce the transgenic mice of the invention. Table 1, below, lists some known viral and cellular oncogenes, many of which are homologous to DNA sequences endogenous to mice and/or humans, as indicated. The term "oncogene" encompasses both the viral sequences and the homologous endogenous sequences.

TABLE 1

Abbreviation	Virus
src	Rous Sarcoma Virus (Chicken)
yes	Y73 Sarcoma Virus (Chicken)
fps	Fujinami (St Feline) Sarcoma Virus (Chicken, Cat)
abl	Abelson Marine Leukemia Virus (Mouse)
ros	Rochester-2 Sarcoma Virus (Chicken)
fgr	Gardner-Rasheed Feline Sarcoma Virus (Cat)
erbB	Avian Erythroblastosis Virus (Chicken)
fms	McDonough Feline Sarcoma Virus (Cat)
mos	Moloney Murine Sarcoma Virus (Mouse)
raf	3611 Murine Sarcoma+ Virus (Mouse)
Ha-ras-1	Harvey Murine Sarcoma Virus (Rat) (Balb/c mouse; 2 loci)
Ki-ras 2	Kirsten Murine Sarcoma Virus (Rat)
Ki-ras 1	Kirsten Murine Sarcoma Virus (Rat)
myc	Avian MC29 Myelocytomatosis Virus (Chicken)
myt	Avian Myelo Blastomas (Chicken)
fos	FBJ Osteosarcoma Virus (Mouse)
ski	Avian SKV T10 Virus (Chicken)
rel	Reticuloendotheliosis Virus (Turkey)
sis	Simian Sarcoma Virus (Woolly Monkey)
N-myc	Neuroblastomas (Human)
N-ras	Neuroblastoma, Leukemia Sarcoma Virus (Human)
Blym	Bursal Lymphomas (Chicken)
mam	Mammary Carcionoma (Human)
neu	Neuro, Glioblastoma (Rat)
ertAl	Chicken AEV (Chicken)
ra-ras	Rasheed Sarcoma Virus

TABLE 1-continued

Abbreviation	Virus
	(Rat)
mnt-myc	Carcinoma Virus MH2 (Chicken)
myc	Myelocytomatosis OK10 (Chicken)
myb-ets	Avian myeloblastosis/ erythroblastosis Virus E26 (Chicken)
raf-2	3611-MSV (Mouse)
raf-1	3611-MSV (Mouse)
Ha-ras-2	Ki-MSV (Rat)
erbB	Erythroblastosis virus (Chicken)

The animals of the invention can be used to test a material suspected of being a carcinogen, by exposing the animal to the material and determining neoplastic growth as an indicator of carcinogenicity. This test can be extremely sensitive because of the propensity of the transgenic animals to develop tumors. This sensitivity will permit suspect materials to be tested in much smaller amounts than the amounts used in current animal carcinogenicity studies, and thus will minimize one source of criticism of current methods, that their validity is questionable because the amounts of the tested material used is greatly in excess of amounts to which humans are likley to be exposed. Furthermore, the animals will be expected to develop tumors much sooner because they already contain an activated oncogene. The animals are also preferable, as a test system, to bacteria (used, e.g., in the Ames test) because they, like humans, are vertebrates, and because carcinogenicity, rather than mutogenicity, is measured.

The animals of the invention can also be used as tester animals for materials, e.g. antioxidants such as beta-carotine or Vitamin E, thought to confer protection against the development of neoplasms. An animal is treated with the material, and a reduced incidence of neoplasm development, compared to untreated animals, is detected as an indication of protection. The method can further include exposing treated and untreated animals to a carcinogen prior to, after, or simultaneously with treatment with the protective material.

The animals of the invention can also be used as a source of cells for cell culture. Cells from the animals may advantageously exhibit desirable properties of both normal and transformed cultured cells; i.e., they will be normal or nearly normal morphologically and physiologically, but can, like cells such as NIH 3T3 cells, be cultured for long, and perhaps indefinite, periods of time. Further, where the promoter sequence controlling transcription of the oncogene sequence is inducible, cell growth rate and other culture characteristics can be controlled by adding or eliminating the inducing factor.

Other features and advantages of the invention will be apparent from the description of the preferred embodiments, and from the claims.

DESCRIPTION OF THE PREFERRED EMBODIMENTS

The drawings will first briefly be described.

DRAWINGS

FIG. 1 is a diagrammatic representation of a region of a plasmid bearing the mouse myc gene and flanking regions.

FIG. 2 is a diagrammatic represenation of a region of a plasmid, pA9, bearing the mouse mammary tumor virus long terminal repeat (MMTV LTR) sequences.

FIGS. 3–6 and 8 are diagrammatic representations of 5 activated oncogene fusions.

FIG. 7 is a diagrammatic representation of a probe useful for detecting activated myc fusions.

MMTV-MYC FUSED GENES

10 Gene fusions were made using the mouse myc gene and the MMTV LTR. The myc gene is known to be an activatable oncogene. (For example, Leder et al. (1983) *Science* 222, 765 explains how chromosomal translocations that characterize Burkitt's Lymphoma and mouse 15 plasmacytomas result in a juxtaposition of the myc gene and one of the immunoglobulin constant regions; amplification of the myc gene has also been observed in transformed cell lines.) FIG. 1 illustrates the subclone of the mouse myc gene which provided the myc regions.

20 The required MMTV functions were provided by the pA9 plasmid (FIG. 2) that demonstrated hormone inducibility of the p21 protein; this plasmid is described in Huang et al. (1981) *Cell* 27, 245. The MMTV functions on pA9 include the region required for glucocorticoid 25 control, the MMTV promoter, and the cap site.

The above plasmids were used to construct the four fusion gene contructions illustrated in FIGS. 3–6. The constructions were made by deleting from pA9 the Sma-EcoRI region that included the p21 protein coding 30 sequences, and replacing it with the four myc regions shown in the Figures. Procedures were the conventional techniques described in Maniatis et al. (1982) *Molecular Cloning: A Laboratory Manual* (Cold Spring Harbor Laboratory). The restriction sites shown in 35 FIG. 1 are StuI (St), SmaI (Sm), EcoRI (R), HindIII (H), PvuI (P), BamHI (B), XbaI (X), and ClaI (C). The solid arrows below the constructions represent the promoter in the MMTV LTR and in the myc gene. The size (in Kb) of the major fragment, produced by diges-40 tion with BamHI and ClaI, that will hybridize to the myc probe, is shown for each construction.

MMTV-H3 myc (FIG. 5) was constructed in two steps: Firstly, the 4.7 Kb Hind III myc fragment which contains most of the myc sequences was made blunt 45 with Klenow polymerase and ligated to the pA9 SmaI-EcoRI vector that had been similarly treated. This construction is missing the normal 3' end of the myc gene. In order to introduce the 3' end of the myc gene, the PvuI-PvuI fragment extending from the middle of the 50 first myc intron to the pBR322 PvuI site in the truncated MMTV-H3 myc was replaced by the related PvuI-PvuI fragment from the mouse myc subclone.

The MMTV-Xba myc construction (FIG. 3) was produced by first digesting the MMTV-Sma myc plas-55 mid with SmaI and XbaI. The XbaI end was then made blunt with Klenow polymerase and the linear molecule recircularized with T4 DNA ligase. The MMTV-Stu myc (FIG. 6) and the MMTV-Sma myc (FIG. 4) constructions were formed by replacing the p21 protein 60 coding sequences with, respectively, the StuI-EcoRI or SmaI-EcoRI myc fragments (the EcoRI site is within the pBR322 sequences of the myc subclone). As shown in FIG. 1, there is only one StuI site within the myc gene. As there is more than one SmaI site within the 65 myc gene (FIG. 4), a partial SmaI digestion was carried out to generate a number of MMTV-Sma myc plasmids; the plasmid illustrated in FIG. 4 was selected as not showing rearrangements and also including a sufficiently long region 5' of the myc promoter (approximately 1 Kb) to include myc proximal controlling regions.

The constructions of FIGS. 4 and 6 contain the two promoters naturally preceding the unactivated myc gene. The contruction of FIG. 5 has lost both myc promoters but retains the cap site of the shorter transcript. The construction of FIG. 3 does not include the first myc exon but does include the entire protein coding sequence. The 3' end of the myc sequence in all of the illustrated constructions is located at the HindIII site approximately 1 Kb 3' to the myc polyA addition site.

These constructions were all checked by multiple restriction enzyme digestions and were free of detectable rearrangements.

PRODUCTION OF TRANSGENIC MICE CONTAINING MMTV-MYC FUSIONS

The above MMTV-myc plasmids were digested with SalI and EcoRI (each of which cleaves once within the pBR322 sequence) and separately injected into the male pronuclei of fertilized one-cell mouse eggs; this resulted in about 500 copies of linearized plasmid per pronucleus. The injected eggs were then transferred to pseudo-pregnant foster females as described in Wagner et al. (1981) *P.N.A.S. U.S.A.* 78, 5016. The eggs were derived from a CD-1 X C57Bl/6J mating. Mice were obtained from the Charles River Laboratories (CD^R-1-Ha/Icr (CD-1), an albino outbred mouse) and Jackson Laboratories (C57Bl/6J), and were houseo in an environmentally controlled facility maintained on a 10 hour dark: 14 hour light cycle. The eggs in the foster females were allowed to develop to term.

ANALYSIS OF TRANSGENIC MICE

At four weeks of age, each pup born was analyzed using DNA taken from the tail in a Southern hybridization, using a ^{32}P DNA probe (labeled by nick-translation). In each case, DNA from the tail was digested with BamHI and ClaI and probed with the ^{32}P-labeled BamHI/HindIII probe from the normal myc gene (FIG. 1).

The DNA for analysis was extracted from 0.1–1.5 cm sections of tail, by the method described in Davis et al. (1980) in Methods in Enzymology, Grossman et al., eds., 65, 404, except that one chloroform extraction was performed prior to ethanol precipitation. The resulting nucleic acid pellet was washed once in 80% ethanol, dried, and resuspended in 300 μl of 1.0 mM Tris, pH 7.4, 0.1 mM EDTA.

Ten μl of the tail DNA preparation (approximately 10 μg DNA) were digested to completion, electrophoresed through 0.8% agarose gels, and transferred to nitrocellulose, as described in Southern (1975) *J. Mol. Biol.* 98, 503. Filters were hybridized overnight to probes in the presence of 10% dextran sulfate and washed twice in 2 X SSC, 0.1% SDS at room temperature and four times in 0.1 X SSC, 0.1% SDS at 64° C.

The Southern hybridizations indicated that ten founder mice had retained an injected MMTV-myc fusion. Two founder animals had integrated the myc gene at two different loci, yielding two genetically distinct lines of transgenic mice. Another mouse yielded two polymorphic forms of the integrated myc gene and thus yielded two genetically distinct offspring, each of which carried a different polymorphic form of the gene.

Thus, the 10 founder animals yielded 13 lines of transgenic offspring.

The founder animals were mated to uninjected animals and DNA of the resulting thirteen lines of transgenic offspring analyzed; this analysis indicated that in every case the injected genes were transmitted through the germline. Eleven of the thirteen lines also expressed the newly acquired MMTV-myc genes in at least one somatic tissue; the tissue in which expression was most prevelant was salivary gland.

Transcription of the newly acquired genes in tissues was determined by extracting RNA from the tissues and assaying the RNA in an Sl nuclease protection procedure, as follows. The excised tissue was rinsed in 5.0 ml cold Hank's buffered saline and total RNA was isolated by the method of Chrigwin et al. (1979) *Biochemistry* 18, 5294, using the CsCl gradient modification. RNA pellets were washed twice by reprecipitation in ethanol and quantitated by absorbance at 260 nm. An appropriate single stranded, uniformly labeled DNA probe was prepared as described by Ley et al. (1982) *PNAS USA* 79, 4775. To test for transcription of the MMTV-Stu myc fusion of FIG. 6, for example, the probe illustrated in FIG. 7 was used. This probe extends from a SmaI site 5' to the first myc exon to an SstI site at the 3' end of the first myc exon. Transcription from the endogenous myc promoters will produce RNA that will protect fragments of the probe 353 and 520 base pairs long; transcription from the MMTV promoter will completely protect the probe and be revealed as a band 942 base pairs long, in the following hybridization procedure.

Labelled, single-stranded probe fragments were isolated on 8M urea 5% acrylamide gels, electroeluted, and hybridized to total RNA in a modification of the procedure of Berk et al. (1977) *Cell* 12, 721. The hybridization mixture contained 50,000 cpm to 100,000 cpm of probe (SA = 10^8 cpm/μg), 10 μg total cellular RNA, 75% formamide, 500 mM NaCl, 20 mM Tris pH 7.5, 1 mM EDTA, as described in Battey et al. (1983) *Cell* 34, 779. Hybridization temperatures were varied according to the GC content in the region of the probe expected to hybridize to mRNA. The hybridizations were terminated by the addition of 1500 units of Sl nuclease (Boehringer Mannheim). Sl nuclease digestions were carried out at 37° C. for 1 hour. The samples were then ethanol-precipitated and electrophoresed on thin 8M urea 5% acrylamide gels.

Northern hybridization analysis was also carried out, as follows. Total RNA was electrophoresed through 1% formaldehyde 0.8% agarose gels, blotted to nitrocellulose filters (Lehrach et al. (1979) *Biochemistry* 16, 4743), and hybridized to nick-translated probes as described in Taub et al. (1982) *PNAS USA* 79, 7837. The tissues analyzed were thymus, pancreas, spleen, kidney, testes, liver, heart, lung, skeletal muscle, brain, salivary gland, and preputial gland.

Both lines of mice which had integrated and were transmitting to the next generation the MMTV-Stu myc fusion (FIG. 6) exhibited transcription of the fusion in salivary gland, but in no other tissue.

One of two lines of mice found to carry the MMTV-Sma myc fusion (FIG. 4) expressed the gene fusion in all tissues examined, with the level of expression being particularly high in salivary gland. The other line expressed the gene fusion only in salivary gland, spleen, testes, lung, brain, and preputial gland.

Four lines of mice carried the MMTV-H3 myc fusion (FIG. 5). In one, the fusion was transcribed in testes, lung, salivary gland, and brain; in a second, the fusion was transcribed only in salivary gland; in a third, the fusion was transcribed in none of the somatic tissues tested; and in a fourth, the fusion was transcribed in salivary gland and intestinal tissue.

In two mouses lines found to carry the MMTV-Xba myc fusion, the fusion was transcribed in testes and salivary gland.

RSV-MYC FUSED GENES

Referring to FIG. 8, the plasmid designated RSV-S107 was generated by inserting the EcoRI fragment of the S107 plasmacytoma myc gene, (Kirsch et al. (1981) *Nature* 293, 585) into a derivative of the Rous Sarcoma Virus (RSV) enhancer-containing plasmid (pRSVcat) described in Gorman et al. (1982) *PNAS USA* 79, 6777, at the EcoRI site 3' to the RSV enhancer sequence, using standard recombinant DNA techniques. All chloramphenicol acetyl transferase and SV40 sequences are replaced in this vector by the myc gene; the RSV promoter sequence is deleted when the EcoRI fragments are replaced, leaving the RSV enhancer otherwise intact. The original translocation of the myc gene in the S107 plasmacytoma deleted the two normal myc promoters as well as a major portion of the untranslated first myc exon, and juxtaposed, 5' to 5', the truncated myc gene next to the α immunoglobulin heavy chain switch sequence.

The illustrated (FIG. 8) regions of plasmid RSV-S107 are: crosshatched, RSV sequences; fine-hatched, alpha 1 coding sequences; left-hatched, immunoglobulin alpha switch sequences; right-hatched, myc exons. The thin lines flanking the RSV-S107 myc exon represent pBR322 sequences. The marked restriction enzyme sites are: R, EcoRI; X, XbaI; P, Pst 1; K, Kpn 1; H, HindIII; B, BamHI. The sequences used for three probes used in assays described herein (C-α, α-sw and c-myc) are marked.

PRODUCTION OF TRANSGENIC MICE

Approximately 500 copies of the RSV-S107 myc plasmid (linearized at the unique Kpn-1 site 3' to the myc gene) were injected into the male pronucleus of eggs derived from a C57BL/6J x CD-1 mating. Mice were obtained from Charles River Laboratories (CD-1, an albino outbred mouse) and from Jackson Laboratories (C57BL/6J). These injected eggs were transferred into pseudopregnant foster females, allowed to develop to term, and at four weeks of age the animals born were tested for retention of the injected sequences by Southern blot analysis of DNA extracted from the tail, as described above. Of 28 mice analyzed, two males were found to have retained the new genes and both subsequently transmitted these sequences through the germline in a ratio consistent with Mendelian inheritance of single locus.

First generation transgenic offspring of each of these founder males were analyzed for expression of the rearranged myc genes by assaying RNA extracted from the major internal tissues and organs in an Sl nuclease protection assay, as described above. The hearts of the offspring of one line showed aberrant myc expression; the other 13 tissues did not.

Backcrossing (to C57Bl/6J) and in-breeding matings produced some transgenic mice which did not demonstrate the same restriction site patterns on Southern blot analysis as either their transgenic siblings or their parents. In the first generation progeny derived from a mating between the founder male and C57BL/6J females, 34 F1 animals were analyzed and of these, 19 inherited the newly introduced gene, a result consistent with the founder being a heterozygote at one locus. However, of the 19 transgenic mice analyzed, there were three qualitatively different patterns with respect to the more minor myc hybridizing fragments.

In order to test the possibility that these heterogenous genotypes arose as a consequence of multiple insertions and/or germline mosacism in the founder, two F1 mice (one carrying the 7.8 and 12 Kb BamHI bands, and the other carrying only the 7.8 Kb BamHI band) were mated and the F2 animals analyzed. One male born to the mating of these two appeared to have sufficient copies of the RSV-S107 myc gene to be considered as a candidate for having inherited the two alleles; this male was backcrossed with a wild-type female. All 23 of 23 backcross offspring analyzed inherited the RSV-S107 myc genes, strongly suggesting that the F2 male mouse had inherited two alleles at one locus. Further, as expected, the high molecular weight fragment (12 Kb) segregated as a single allele.

To determine whether, in addition to the polymorphisms arising at the DNA level, the level of aberrant myc expression was also altered, heart mRNA was analyzed in eight animals derived from the mating of the above double heterozygote to a wild-type female. All eight exhibited elevated myc mRNA, with the amount appearing to vary between animals; the lower levels of expression segregated with the presence of the 12 Kb myc hybridizing band. The level of myc mRNA in the hearts of transgenic mice in a second backcross generation also varied. An F1 female was backcrossed to a C57Bl/6J male to produce a litter of seven pups, six of which inherited the RSV-S107 myc genes. All seven of these mice were analyzed for expression. Three of the six transgenic mice had elevated levels of myc mRNA in the hearts whereas in the other three the level of myc mRNA in the hearts was indistinguishable from the one mouse that did not carry the RSV-S107 myc gene. This result suggests that in addition to the one polymorphic RSV-S107 myc locus from which high levels of heart-restricted myc mRNA were transcribed, there may have been another segregating RSV-S107 myc locus that was transcriptionally silent.

CARCINOGENICITY TESTING

The animals of the invention can be used to test a material suspected of being a carcinogen, as follows. If the animals are to be used to test materials thought to be only weakly carcinogenic, the transgenic mice most susceptible of developing tumors are selected, by exposing the mice to a low dosage of a known carcinogen and selecting those which first develop tumors. The selected animals and their descendants are used as test animals by exposing them to the material suspected of being a carcinogen and determining neoplastic growth as an indicator of carcinogenicity. Less sensitive animals are used to test more strongly carcinogenic materials. Animals of the desired sensitivity can be selected by varying the type and concentration of known carcinogen used in the selection process. When extreme sensitivity is desired, the selected test mice can consist of those which spontaneously develop tumors.

TESTING FOR CANCER PROTECTION

The animals of the invention can be used to test materials for the ability to confer protection against the development of neoplasms. An animal is treated with the material, in parallel with an untreated control transgenic animal. A comparatively lower incidence of neoplasm development in the treated animal is detected as an indication of protection.

TISSUE CULTURE

The transgenic animals of the invention can be used as a source of cells for cell culture. Tissues of transgenic mice are analyzed for the presence of the activated oncogene, either by directly analyzing DNA or RNA, or by assaying the tissue for the protein expressed by the gene. Cells of tissues carrying the gene can be cultured, using standard tissue culture techniques, and used, e.g., to study the functioning of cells from normally difficult to culture tissues such as heart tissue.

DEPOSITS

Plasmids bearing the fusion genes shown in FIGS. 3, 4, 5, 6, and 8 have been deposited in the American Type Culture Collection, Rockville, Md., and given, respectively, ATCC Accession Nos. 39745, 39746, 39747, 39748, and 39749.

OTHER EMBODIMENTS

Other embodiments are within the following claims. For example, any species of transgenic animal can be employed. In some circumstances, for instance, it may be desirable to use a species, e.g., a primate such as the rhesus monkey, which is evolutionarily closer to humans than mice.

We claim:

1. A transgenic non-human mammal all of whose germ cells and somatic cells contain a recombinant activated oncogene sequence introduced into said mammal, or an ancestor of said mammal, at an embryonic stage.

2. The mammal of claim 1, a chromosome of said mammal including an endogenous coding sequence substantially the same as a coding sequence of said oncogene sequence.

3. The mammal of claim 2, said oncogene sequence being integrated into a chromosome of said mammal at a site different from the location of said endogenous coding sequence.

4. The mammal of claim 2 wherein transcription of said oncogene sequence is under the control of a promoter sequence different from the promoter sequence controlling the transcription of said endogenous coding sequence.

5. The mammal of claim 4 wherein said promoter sequence controlling transcription of said oncogene sequence is inducible.

6. The mammal of claim 1 wherein said oncogene sequence comprises a coding sequence of a c-myc gene.

7. The mammal of claim 1 wherein transcription of said oncogene sequence is under the control of a viral promoter sequence.

8. The mammal of claim 7 wherein said viral promoter sequence comprises a sequence of an MMTV promoter.

9. The mammal of claim 7 wherein said viral promoter sequence comprises a sequence of an RSV promoter.

10. The mammal of claim 1 wherein transcription of said oncogene sequence is under the control of a synthetic promoter sequence.

11. The mammal of claim 1, said mammal being a rodent.

12. The mammal of claim 11, said rodent being a mouse.

* * * * *

5 10 15 20 25 30 35 40 45 50 55 60 65

Index